百工探秘

HOW IT WORKS

（上卷）

赵致真　张戟○主编　[德]红箭○著　高洁　芒冰○译

长江出版传媒　湖北科学技术出版社

图书在版编目（ＣＩＰ）数据

百工探秘（上卷）／ 赵致真，张戟主编．［德］红箭(Red Arrow)著．
高洁，芒冰译-- 武汉 ：湖北科学技术出版社，2021.7
ISBN 978-7-5706-1513-1

Ⅰ．①百… Ⅱ．①赵… ②张… ③德… ④高… ⑤芒…
Ⅲ．①产品设计－介绍－世界 Ⅳ．①TB472

中国版本图书馆CIP数据核字(2021)第 081110 号

百工探秘（上卷）
BAIGONG TANMI （SHANGJUAN）

特约编辑：于雯雯
责任编辑：彭永东   封面设计：胡　　博

出版发行：湖北科学技术出版社   电话：027-87679468
地　　址：武汉市雄楚大街 268 号   邮编：430070
　　　　　（湖北出版文化城 B 座 13-14 层）

网　　址：http://www.hbstp.com.cn

印　　刷：武汉精一佳印刷有限公司   邮编：430034

787×1092　　　　　1/16　　　　　　　　20 印张　　　　400 千字
2021 年 7 月第 1 版　　　　　　　　　2021 年 7 月第 1 次印刷
　　　　　　　　　　　　　　　　　　　　（上下卷）定价：198.00 元

2006 年北京科教电视节开幕那天，有位外国评委小心翼翼地送给我一个打火机。稠人广众之下，我只是礼貌性地点头致谢。由于自己不抽烟，我便转手送给了大会的司机。事后有人告诉我，这是一款"芝宝打火机"，世界名牌。我听了仍然没有感觉。

直到前几年，《科技之光》从德国引进了一套系列电视片《百工探秘》，我饶有兴致地审看了 160 集全部内容，其中正好有一集"芝宝打火机"，这才让我大开眼界。原来"芝宝打火机"是好莱坞电影中多次出现的醒目道具，钢片制成的风挡上有 16 个圆孔，能让火焰在大风中不被吹熄。生产线上先后有 20 个戴着"米老鼠手套"的质检员严格把关，独特的"咔嗒"声，是它一成不变的品牌标记。终生保修和以旧换新，则是它信守不易的服务承诺。这集 5 分钟的电视片让我看得肃然起敬又惶愧不安，很自然想到当年北京科教电视节，在送我"芝宝打火机"的外国朋友面前，竟然没有说出一句"内行话"，该让人家多么失望和扫兴。他不一定知道汉语中"孤陋寡闻"和"明珠暗投"的成语，但没准会摇头叹息一声"不识货"和"没文化"吧？

当后来策划出版一组"融媒体图书"时，我们便首先想到了《百工探秘》。而单独引进国外电视片的"图书播放权"，这在国内还没有先例。

《百工探秘》的读者，首先是广大的"你我他"。上下两卷所采纳的 150 集的巨大容量，扫二维码观看中文、英文视频，无疑是瞭望世界物质文明的一扇窗口，也是涉猎天下品牌的一次巡礼。我过去真不知道，仅欧洲就制定了这么多专门的法律条款，来监督和保护自己的"驰名商标"：苏格兰威士忌酒必须装在橡木桶中按年头开酿；德国啤酒生产中只能使用水、大麦、啤酒花和酵母；帕马森奶酪仅限于意大利特定地区生产；哥斯达黎加出口欧洲的香蕉，无论大小、形状、曲线都务要中规中矩。如果企业稍有疏失，就直接"违法"了。

《百工探秘》的覆盖之广，描述之细，拍摄之精，都是令人赞服的。取材看上去百业杂陈，但每集内容都各有千秋。

从比利时巧克力、日本酱油、加拿大枫糖浆，到美国大米、伊朗开心果、南非牛油果；从波士西装、开司米羊绒、威灵顿雨靴，到三角钢琴、乐高积木、模型赛车；从斯沃琪手表、国际标准足球、哈雷摩托车，到风力光伏发电、摩天大楼建造、地下金矿开采。这里登台亮相的厂商们或是全球行业中的一方翘楚，或是所在国财政收入的重要来源，或是植根于古老传统中的民族文化遗产。"世事洞明皆学问"。了解并欣赏这些殊方异域的七行八作和物华天宝，对于长知识、广见闻、通人情、晓社会，形成更完整和深厚的"天下观"，应该是大有裨益的。

《百工探秘》的读者，更应该是少年朋友。大约没有哪一代孩子，像今天这样丰衣足食和用度阔绰，享尽了现代社会的便利舒适。但若问他们身边的各种东西从何而来？恐怕很多人只能回答：来自超市的货架、网上的购物、快递的包裹。而对于制造这些产品的过程则大都毫无所知。

《百工探秘》描述了衣食住行各方面诸多产品的设计理念和生产流程，并回答了孩子们最爱问的许多"为什么"：为什么红葡萄酒杯"肚子"鼓起而白葡萄酒杯"腰身"细长？为什么登山鞋的面料防水却又能排汗透气？为什么铝制行李箱的外壳做成瓦楞形？为什么暖气片要安装在窗户下？对于我们目前教育的"短板"，这本书也许不无小补。

时常从媒体上看到能歌善舞的娱乐神童，博闻强记的诗词奇才，成为孩子惊羡不已的人生偶像。其实我们还有许多痴迷工程技术的少年才俊，也需要成长的环境和土壤。《百工探秘》并非直接讲科学，但处处都是科学向技术的转化。材料的学问、机械的原理、效率的实现、质量的控制、流水线的衔接，无疑能让孩子从小亲近生产劳动环境，习惯"数理化思维"，培养工程师头脑。中国新一代的"能工巧匠"也应该"从娃娃抓起"。

《百工探秘》的读者，最应该是企业家和工程技术人员。在"百年未有的大变局"下，我们如何确保"制造大国"和"世界工厂"的优势地位？如何提升竞争力并占领更大份额的全球市场？如何从跟跑，并跑，转变为全面领跑？毫无疑问，知己知彼是胜人一筹的前提条件。学习他人的寸长片善，则是超越对手的终南捷径。

据说世界渐渐兴起了"工业旅游热"。但许多地方都是"生产重地，闲人免进"的。即使来到厂房车间，也会处处看到"禁止拍照"的赫然警示。任何企业都会格外注重保护"商业机密"。在知识产权神圣不可侵犯的当今时代，一切通过"逾墙窥隙"获取"科技情报"的手段，都为法律所不容，更为诚信企业家所不齿。那么该如何"善于学习"呢？其实大量有价值的资讯，恰恰都可以从公开的媒体报道中获取。《百工探秘》的一个"探"字，肯定不是"刺探"和"暗探"，而是"探寻"和"探求"。然而，当《百工探秘》的高清摄像机拍下一条条流水线的生动细节和特写画面时，"外

行看热闹，内行看门道"，究竟会触发谁的"创意点子"和"头脑风暴"，就看个人的悟性和造化了。

《百工探秘》的"探"字，也许还在"探讨"另一种"奥秘"。我们曾经见到过不少品牌，红极一时却朝荣夕悴。而《百工探秘》中的诸多世界名牌却驰声走誉，百年不息。其中的原因究竟是什么？广告效应自然不能忽视，但"巧伪不如拙诚，表壮不如里壮"，任何包装和炒作都不能取代产品质量；生产手段自然十分关键，但无论如何现代化、智能化，都不能取代工匠精神。《百工探秘》的每一集短片，如同生动美丽的乐句和乐段，共同唱响了劳动创造的颂歌。

我们编写这本《百工探秘》的初衷可以一言蔽之：借"他山之石"砥节砺行。让品牌意识和工匠精神成为"中国制造"的灵魂！

目 录
CONTENTS

美食篇

比利时巧克力

比利时首都布鲁塞尔除了世界著名的喷泉还有一个大概更有品位的标志：巧克力。它具有不同形状和大小。下面讲讲比利时工厂是如何大规模生产巧克力的。

这辆卡车中装载的，是本节目的明星——纯而又纯的巧克力。为了让卡车罐中的巧克力保持柔软，温度被控制在47摄氏度。巧克力从卡车里抽进工厂的储存罐。

每一天，这家工厂生产50多万颗巧克力糖和松露巧克力，这些储存罐都盛满了巧克力。巧克力被用于制作数百种不同品种的糖果，这个国家最受欢迎的一款是卡普利斯巧克力。成袋的榛子和杏仁被用来制造外壳。它们在115摄氏度下烤15分钟。大量的玉米糖浆放进碗里加入糖粉。混合物加热后会变成焦糖。此时烤坚果已被磨碎，可以加进混合物中了。混合物被放到传送带上并形成果仁脆糖。

切成方块送到加热的台面，那里的工人把它卷成卡普利斯巧克力形状。卡普利斯巧克力的馅是香草奶油，对注意自己体重的人不太适合。加入大块的黄油、糖，然后是香草精、卵磷脂，最后为了保证你的医生不会认可，再加入一团白巧克力。混合物被送到生产线上，工人单独填充每个糖卷。

卡普利斯巧克力糖的收尾工序，当然是一道华丽的巧克力涂层。

由于要保持比利时巧克力在全世界的声誉，巧克力粉被定期检查，有人制作了一台机器来做这件事。甚至巧克力的温度也是很重要的，因为它会影响糖果最终的味道和质地。

除了经典糖果比如卡普利斯巧克力之外，这个工厂会定期增加新品种。一个新的松露巧克力，从笔记本上的勾画开始。要成为巧克力设计师，你不必是一个伟大的艺术家，只需要有丰富的想象力。

在工厂的厨房里对多种成分进行测试，以得到完美的味道和外观。当他对巧克力的质地和口感表示满意后，生产商可以开始制作新的松露巧克力样品了。他舀进牛奶巧克力的混合物，让它固化。当托盘翻过来时，尚未固化的巧克力掉出来，留下一层巧克力壳。然后，他在覆盖另一层巧克力之前灌进填充物。当它们硬化后就可以被敲打出来。新的配方得到了批准，可以进行大规模生产了。

就像此前看到的，巧克力被泵入托盘。当巧克力部分固化时，托盘被翻转过来，留下一层巧克力壳。巨大的电扇帮助巧克力壳变硬。然后泵入填充物。最后添加的一层巧克力会成为基底层。当然，多余的巧克力会被重复使用。轻微扭转托盘，能使硬化了的巧克力松动。巧克力从生产线出来时上下颠倒，简单地翻转过来就行。它作为一个新加冕的竞争者，已经准备好夺取布鲁塞尔的巧克力王位了。

你知道吗？

巧克力中有一种能让人产生欣快感的化合物，作用和咖啡因类似，它叫作可可碱——是一种能让狗生病的化学物质。

1.巧克力是比利时首都布鲁塞尔的标志
2.柔软的巧克力从卡车罐中抽进储存罐
3.制作卡普利斯巧克力，要先把烤熟磨碎的坚果放入焦糖搅拌

4.混合物被切成方块送至加热台面
5.将白巧克力、黄油、香草精、卵磷脂搅拌成香草奶油
6.香草奶油填充到每个糖卷中

7. 收尾工序是加巧克力涂层

8. 专用机器对巧克力粉定期检查

9. 新品种先制作样品

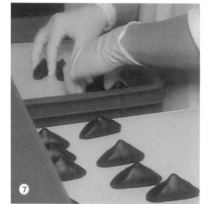

10. 样品制作流程

11. 新配方投入生产线

12. 新款产品准备竞争布鲁塞尔的
巧克力王冠

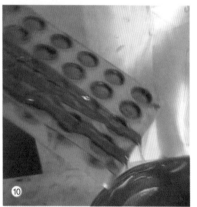

Belgian Chocolate

扫描二维码，观看英文视频。

Brussels: as well as a world famous fountain the Belgian capital has a trademark that's arguably more tasteful: chocolate and it comes in all shapes and sizes. This is how a Belgian factory makes chocolates on a massive scale.

In this truck is the star of the show. Pure unadulterated chocolate. To keep it soft in the trucks tanks the chocolate is kept at forty seven degrees Celsius. From there it's pumped into the factory's storage tanks.

Every single day this factory produces over half a million individual chocolates and truffles and these tanks are kept topped up with chocolate. It will be used for hundreds of different varieties of sweets and one of the nations favourite is the Caprice. Sacks of hazelnuts and almonds are used to make the shell. They're roasted for 15 minutes at one hundred and fifteen degrees Celsius. Large quantities of corn syrup are put into a bowl and combined with icing sugar. When it's heated the mix caramelises. Meanwhile the roast nuts have been ground down and now they can be added to the mix. The mixture is laid out on to this conveyor belt and sets to form a nut brittle.

It's cut into squares and passed along onto a heated surface, where workers roll them into the Caprice shape. The filling for the Caprice is a rich vanilla cream and this is not the weight watchers variety. Blocks of butter and sugar are followed by vanilla extract, lecithin to bind it all together and finally just to make absolutely certain your doctor wouldn't approve a dollop of white chocolate. The mix is taken to the production line where the workers fill each roll individually.

The finishing touch of a caprice is of course a sumptuous chocolate coating.

Because Belgian chocolate has a world-wide reputation to maintain the chocolate mix is regularly checked and somebody created a machine to do it. Even the temperature of the chocolate is important as it affects taste and texture of the final sweets.

As well as golden oldies like the Caprice this factory regularly adds new varieties to it's range. A new chocolate truffle begins life as an idea sketched onto a pad. You don't have to be a great artist to be a chocolate designer all you need is a good imagination.

In the in-house kitchen a variety of ingredients are tested to get the taste and look perfect. Once he is happy with the texture and taste the chocolatier can begin crafting the first samples of his new chocolate truffle. He spoons in a milk chocolate mix and leaves it till it starts to set. When he turns the tray over the soft centre that hasn't yet set falls out leaving chocolate shells. He then pipes in the filling before covering the base with another layer of chocolate. When they've hardened they can be tapped out. The new recipe approved and it goes in to full scale production.

Just like before chocolates pumped into the trays. And when it's partly set the trays are tipped over leaving behind the chocolate shells. Giant fans cool the chocolate to encourage the shell to harden. The filling is then pumped in. A final layer of chocolate makes the base. Of course the excess is all reused. A slight twist loosens the hardened sweets from their tray. They come down the line upside down so they're simply flipped over, And it's ready a newly crowned contender competing for the Brussels throne.

Did you know?

The compound in chocolate that gives human a caffeine-like buzz is called theobromine-a chemical that makes dogs sick.

巧克力蛋糕

我们都爱蛋糕，尤其是巧克力蛋糕。这是世界上最奢华的巧克力蛋糕之一——维也纳萨赫巧克力蛋糕。

众所周知做蛋糕需要打几个鸡蛋。有位女士对此了解得更多。她每年要打破百万个以上的鸡蛋，并且将蛋白和蛋黄分离。蛋白被反复搅打直到形成松软的峰尖。

下一个搅拌器里转动的是蛋糕材料。这可算不上健康食品，除了蛋黄和面粉外，还有大量的黄油、巧克力和糖。大桶里的蛋糕材料被倒入数十个直径为21厘米的蛋糕盘，准备拿去烘烤，就像你家里做蛋糕一样。

但在这里，他们每次烤180个蛋糕。在190摄氏度的温度下烤1小时。如果你好几年没有用过烤箱的话，这是烤箱温度调控器的第五档。

蛋糕烤好后，两人操作的流水线要尽快把蛋糕从烤盘里取出来。如果动作慢了，蛋糕会和烤盘粘住而报废。这台机器用一根细金属线，每次把蛋糕横向切成完美的上下两层。给这些蛋糕赋予独特口味的，是蛋糕的夹心。一桶桶甜而黏稠的杏子酱在这些锅里加热，然后被大勺浇到下层的蛋糕上。

当甜点师随意将这些杏子酱浇到蛋糕上时，看起来似乎乱糟糟，但这不成问题，因为整个蛋糕最后会被涂上黏稠的巧克力酱。

任何节食者看到这里，请把目光转向别处。现在要涂蛋糕的最后一层了——大量的黑巧克力酱。蛋糕被放置冷却，然后就可以包装了。如同蛋糕自身的所有配料一样，只有最好才算够好。这意味着一个优雅的木头盒子。

在制作中倾注了爱、关怀和高额卡路里的维也纳萨赫巧克力蛋糕，将注定不会在盒子里待得太久。

你知道吗？

在罗马时代，人们把婚礼蛋糕扔在新娘头上，以祝福新婚夫妇好运和多子。

1. 维也纳萨赫巧克力蛋糕
2. 这里每年打破百万个以上的鸡蛋
3. 分离出的蛋白被反复搅打
4. 把搅拌器里的材料倒入蛋糕盘
5. 送到 190 摄氏度的烘箱烤 1 小时
6. 烤好的蛋糕从烤盘里取出

7.用一根金属线将蛋糕切成上下两层

8.将杏子酱加热后浇到下层蛋糕上

9.整个蛋糕涂上巧克力酱

10.覆盖一层黑巧克力

11.蛋糕被放置冷却

12.巧克力蛋糕装入木制盒子

扫描二维码, 观看英文视频。

Luxury Chocolate Cake

We all love cake, especially chocolate cake. And this is one of the most luxurious chocolate cakes in the world, the Sacher Viennese chocolate torte.

Everyone knows you can't make a cake without breaking a few eggs, and here's a lady who knows better than most, she helps to break and separate over a million eggs every year. The whites are beaten until they form soft peaks.

In the next mixer is the cake mix. This is not health food, along with the yolks and flour there is a generous dose of butter, chocolate and sugar. The vat gets emptied into dozens of 21cm cake tins ready to be baked, just like you might do at home.

But here they bake 180 cakes at a time. They go into the oven for an hour at 190 degrees. That's gas mark 5 if you haven't brought an oven for a few years.

When they are baked a two man production line gets the cakes out of the tins as quickly as possible. If they're left too long the mix will stick to the tin and the cakes will be ruined. This machine slices the cakes with a thin wire leaving two perfect halves every time. It's the filling that's gives these cakes their distinct taste. Barrels of sweet and sticky apricot jam are heated in these pots and then spooned on to the base.

It's looking like a bit of a mess as the confectioner dollops it on with abandon, but that's not a problem as the whole cake is about to be smeared all over with the gooey sauce.

Any dieters who have made it this far, look away now. It's time for the final layer of the cake – lashings of dark chocolate sauce. They're left to cool and then they're ready to be packed. Like all of the ingredients that go in to the cake itself– only the finest is good enough, and that means an elegant wooden box.

The love, care and sheer number of calories that go into making a Sacher Viennese chocolate torte will ensure that it won't stay in the box for long.

Did you know?

In Roman times, wedding cakes were broken over the head of the bride to wish the newlyweds good fortune and fertility.

冰激凌

多数人都喜爱它，尤其在大热天。我们即将揭示这种甜食美味背后的香甜奥秘。我们将探究冰棍，还有冰激凌表面松脆的巧克力涂层工艺。

数据显示 2006 年英国人消费了 2 亿多个手持冰激凌，因此厂商要做大量工作以跟上需求。这意味着以卡车计量的白糖还有液态氮。工人在零下 190 摄氏度操作时，戴手套至关重要。液氮能冻结冰激凌。

做冰激凌和冰棍的第一步始于厨房。多数的大冰激凌厂设有研发部门来创造新的口味、形状和品种。你能想到的所有原料，比如牛奶、奶油和糖都在这里被使用，还有大量的可可粉以及坚果、水果和不寻常的装点物。在冰激凌无休止的创新过程中这些成分的组合都被试验、测试和品尝过。主厨对配方感到满意后车间人员就可以开始工作了。

他们在为今天的生产混合巧克力。尽管整个工厂是自动化的，控制室仍时刻需要管理。技术人员持续监控机器、温度和重要的产出参数。

生产过程中一种最重要成分是不能吃的——用手拿冰激凌的木棍，它由榉木制成。要做成巧克力涂层冰激凌，首先由机器生产长条状香草冰激凌。这个阶段混合物相当软，对插入木棍至关重要。冰激凌块送到底部进行切割，在最后一分钟插入木棍。显然，如果冰激凌此时冻结了，就插不进木棍。接着将木棍放正，不仅使冰激凌外表美观，还便于稍后浸泡到巧克力中。为保持木棍的位置，冰激凌块被送到 -40 摄氏度的巨大冰柜中。这使得冰激凌变硬，把木棍固定。

现在可以涂上著名的涂层了。笔直的木棍将整个冰激凌都蘸进去，涂上一层厚厚的巧克力。很多人担心冰激凌会使人发胖，但这种特殊冰激凌的热量高达 261 卡路里（1 卡 ≈ 4.19 焦耳）却并没有影响人们的喜爱。

巧克力层被咬后特殊的破碎感是冰激凌吸引人的一个重要因素。微小喷嘴的吹风使巧克力硬化成松脆的外层。

冰激凌涂有各种不同的巧克力以满足不同的口味，里面均匀的微小气泡让冰激凌有一种奶油的质感。生产过程中奶油混合物被泵满气泡。冰激凌中的脂肪分子包围气泡防止它们破裂使得冰激凌具有松软的质感。在工厂测试部门工人继续评估自己的产品，以确保达到标准。

市场上有多种味道和品类的冰棒。果味品种的冰棒很受欢迎，暑天里普通冰棒也能给人们带来清凉。不同于巧克力脆皮冰激凌，冰棒的生产简单得多。把水果味的液体倒入模具。插上一根木棍，让它冻结到位。冻实以后，机器抓住木棍将它取出来。然后在清水中蘸一下除去冰晶，这样冰棒就不会粘住包装纸。常常有冰棒达不到标准，但逃不过质检员专注的眼睛。最后冰激凌全部按照传统包装，发送到全国各地商店。对付酷热夏天的良方，就是木棒上带着冷霜的冰激凌。

你知道吗？

在东京的一次冰激凌博览会上，游客们可以品尝到各种奇异的口味，包括鱼翅、大蒜，甚至牛舌冰糕。

1. 制作冰激凌的白糖运到工厂
2. 冻结冰激凌使用零下190摄氏度的液氮
3. 试验不同口味的冰激凌和冰棍
4. 冰激凌生产中加入配料

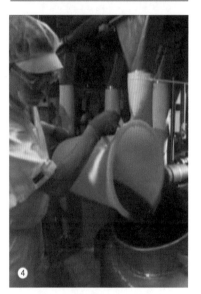

5. 自动化监控生产流程
6. 冰激凌块进行切割并插入木棍

7. 给完成冷冻的冰激凌涂上一层巧克力
8. 微小喷嘴吹风使巧克力形成松脆的外壳
9. 巧克力层有特殊的破碎口感
10. 冰棒生产中将果味液体倒入模具
11. 插上木棍进行冷冻
12. 冰棒通过质检后包装出厂

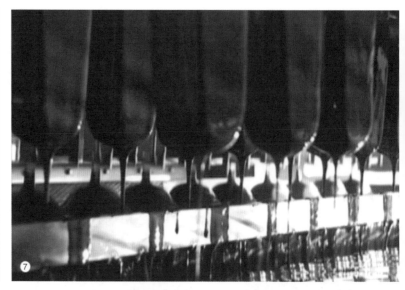

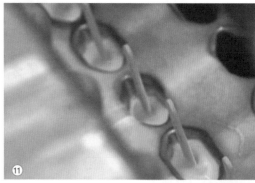

Ice Creams

Most of us love them, especially on a hot day. We're about to reveal the sweet secrets behind the snack that has others licked. We're investigating ice lollies, and the craft that creates that crunchy layer on a chocolate-coated ice cream.

Figures show that in 2006 Briton's consumed more than 200 million hand held ice creams so the manufacturers have a lot of work to do to keep up with demand. That means truckloads of sugar and truckloads of liquid nitrogen too. At minus 190 degrees Celsius it's vital that workers wear gloves when handling it. This is what helps freeze your ice cream.

The first stages of making new ice creams and ice lollies starts in a kitchen like this one. Most big ice cream factories have a development department to invent new flavours, shapes and varieties. All the basics you'd expect like milk, cream and sugar are used here. There's also plenty of cocoa powder as well as nuts, fruits and unusual sprinkles. Combinations of these ingredients are tried, tested and tasted in the unending process of ice cream innovation. Once the chefs are satisfied with their recipe the staff on the factory floor can get to work.

Here they are mixing together the chocolate for today's batch. Although the entire factory is automated, and the control room managed at all times. Technicians constantly monitor machines, temperatures and that all-important output.

One of the most important ingredients in this process isn't actually edible. The wooden sticks used to hold the ice cream with are made of beech. To assemble chocolate coated ice creams, first machines form a long tube of vanilla. The mixture is quite soft at this stage which is vital for adding in the stick. As the block passes to the bottom to be cut, a stick is inserted at the last minute. Obviously if the dessert was frozen solid at this point, you couldn't get the stick in. Next, the sticks are straightened which helps with both the look of the ice cream and for when the block is dunked in chocolate later on. Then to keep the stick in place, the blocks are sent to a huge freezer which chills them to -40 degrees Celsius. This hardens them and locks the stick into place.

And now they're ready for their well-known coating. The straight stick means the whole ice cream is dipped, fully coating it with a thick chocolaty layer. Many people worry that ice cream is fattening, but the whopping 261 calories found in this particular ice cream hasn't stopped it from being very popular.

The characteristic crack of the chocolate breaking-bite is an important part of this ice cream's appeal. To chocolate is hardened to a crunchy coating air blown from tiny nozzles.

The ice creams are coated with different kinds of chocolate to suit different tastes, but on the inside, it's the same tiny bubbles that give the ice cream that creamy texture. During production the creamy mixture is pumped full of bubbles. Fat molecules in the ice cream surround the bubbles and stop them from bursting. And this helps to give the ice cream a lighter fluffier texture. Here in the factory's testing department, staff continually assesses their products to be sure they're up to standard.

A huge variety of ice lolly flavours and types are on the market. Fruity sorbet varieties are popular and ordinary ice lollies are just as refreshing on a hot day.Unlike the chocolate-covered ice cream, an ice lolly is far simpler to produce. Fruit flavoured liquid is poured into a mould. A stick is inserted and it's frozen in place. Once its solid machines grab the stick and it's pulled free. It's then dipped in plain water which removes ice crystals so the lollies don't stick to their wrappers. From time to time, a lolly isn't up to scratch, but it doesn't get past the watchful eyes of the quality control inspector. Finally the ice creams are all wrapped in their traditional packaging and sent out to stores all over the country. So, what's the cure for a meltingly hot summer's day? A frosty cold ice cream on a stick.

Did you know?

At an ice cream trade fair in Tokyo, visitors could try a range of bizarre flavours including shark fin, garlic, and even a cow–tongue sorbet.

酱油

酱油在中餐和日餐里，都是一个关键成分。也在英国超市货架上随处可见。这个日本工厂，拥有几百年的酿造经验，生产被当地人称为"shoyu"，而被我们叫作酱油的产品。酱油的生产从非常简单的原料开始：大豆、小麦，以及一种特殊的霉菌。

负责生产过程的平山师傅要做的第一件事就是确保原料混合比例正确。

他们蒸完大豆，磨好小麦，这时的粗糙混合物被称作"Koji"。霉菌成倍繁殖产生酵素，由此启动发酵过程。在这个房间里，"Koji"被不停翻动，以确保原料充分混合。加进海盐和水，然后将这些湿润的糊状物转移到巨大的罐子中。并在那里放置大约6个月熟化。

此时，这些烂泥一样的东西绝不会让人有胃口，但是从微观层面，变化正在发生。霉菌将大豆蛋白分解成氨基酸。同时，小麦发酵产生糖分，让这个糊状物微带甘甜。这两个反应过程叠加起来，造就了酱油特有的芳香、滋味和颜色。虽然这时候的产品样子难看，但平山师傅却对它的味道非常满意，所以糊状物可以进入下一道工序。这次需要几千米长的亚麻布。

糊状物被均匀地铺到亚麻布上，此后亚麻布被折叠600次，做成一个巨型的大豆三明治。

一队工人把这座"亚麻布塔"拖到液压机上。液压机以200吨的重量挤压亚麻布，榨出棕色的液体。每一座"亚麻布塔"可以产出5000升的生酱油。它被极高的温度加热后，6个月的生产流程就结束了，可以开始装瓶。和前期生产步骤迥然不同的是，装瓶只需要几秒钟时间。这家工厂每年生产2.7亿升的酱油，运送到世界各地的餐盘和消费者口中。

你知道吗?

酱油独特的风味，被称作"鲜"，是甜、酸、咸、苦和鲜五种基本口味之一。

1.酱油对于中餐和日餐都不可或缺

2.酱油原料包括大豆、小麦和一种特殊霉菌

3.蒸好的大豆，磨好的小麦和霉菌按比例混合

4.原料被不停翻动至充分混合

5.加进海盐和水后的糊状物被转移到罐中

6.经 6 个月的熟化后检查发酵状况

7. 霉菌将大豆蛋白分解成氨基酸，小麦发酵产生糖
8. 糊状物均匀地铺到亚麻布上
9. 将折叠600次的亚麻布推上液压机
10. 在200吨重压下酱油从"亚麻布塔"渗出
11. 对生酱油高温灭菌
12. 成品酱油灌装工序

Soy Sauce

Soy sauce is a key ingredient in both Chinese and Japanese food and has become a common sight on British supermarket shelves. In this Japanese factory they use hundreds of years of brewing experience to produce what the locals call 'shoyu' but is best known to the rest of us as soy sauce. It's a sauce that begins with very simple ingredients: soy beans, wheat and a special mould.

Master Hirayama oversees the production process and the first thing is to make sure that they've got the right mix of ingredients.

When they have steamed the beans and milled the wheat they call the lumpy mix "Koji". The mould multiples and produces enzymes and this kick starts the fermentation. In this room the Koji is kept on the move to make sure it's thoroughly mixed. Sea salt and water are added and then the moist mash is transported into huge tanks. It's left there to ripen for about six months.

The sludge doesn't look like anything you would want to eat at the moment but at a microscopic level a transformation is taking place. The mould breaks down the soy bean proteins into amino acids. Meanwhile the wheat ferments and produces sugar giving the mash a slightly sweet taste. Together these two processes give soy sauce it's characteristic aroma, flavour and colour. Despite it's appearance the Master is impressed with flavour so the mash can go on to the next step. This stage requires linen several kilometres of it. The mash is spread evenly on top and then the linen is folded six hundred times to make a massive soy sandwich.

A team of workers haul the tower of linen through to a hydraulic press. The press wrings out the linen with a two hundred ton weight and squeezes out a brown liquid. Each pile yields around 5000 liters of raw soy sauce. It is heated at an ultra high temperature and then the six month process is over and they can bottle it up. In sharp contrast to the rest of the production this is done in just a few seconds. Each year this factory produces two hundred and seventy million liters of soy sauce which will find its way onto the plates and palates of consumers across the planet.

Did you know?

The distinct savoury flavour of soy sauce, called "umami" is one of the 5 basic tastes—sweet, sour, salty, bitter and umami.

奶油奶酪

扫描二维码，观看中文视频。

你会把酸了的牛奶涂到面包上吗？可能不会，但其实奶油奶酪就是这么回事。在英国，奶油奶酪虽然不如车达奶酪那么流行，但它正迅速吃香。

在这家德国工厂，每天要处理45万升牛奶。在把牛奶变成奶油奶酪之前要冷却。接下来牛奶被分离成脱脂奶和奶油。它们此后会按不同数量重新混合，生产出一系列不同脂肪含量的奶油奶酪。

乳品混合物需要进行巴氏消毒。被加热到73摄氏度并持续24秒，以杀灭病菌。此后加入乳酸。当混合物被加热时，乳酸融合到牛奶里，使它更加黏稠并且变酸。

混合物被倒入一个分离器，用来去除多余的液体，称为乳清。如果你还记得《小玛菲特小姐》这首歌，你也许能猜到，剩下来的是凝乳。

奶酪就是由此做出来的，但还需要加点盐。或者说加一小桶盐。此后再次被加热，以防还有细菌存活。

当你想到奶油奶酪时，可能会联想到烟熏三文鱼和贝果，但这家工厂迎合更多口味。奶油奶酪将与干洋葱和干香葱混合。如果你晚上有约会的话，午餐吃这个就需要三思了。

奶油奶酪和其他奶酪不同，不需要熟化的过程，而是趁新鲜吃。虽然奶油奶酪在欧洲广受欢迎，却是在美国被发明的。1872年，一个纽约农民生产出一种美味空前的奶酪。这就是后来广为人知的奶油奶酪。最后一步是用氮气搅拌奶酪使它变得蓬松，搅拌使奶酪体积增加更容易涂抹。

装奶酪的小盒子由这两卷塑料制成。塑料被加热，然后由一台机器将它冲压成型。接下来，每个塑料盒里放进精确的分量。包括低脂瘦身型品种。盒子被密封好，冷却12小时，就可以在超市出售了。

你知道吗？

对奶酪的非理性惧怕被称为"奶酪恐惧症"。

1. 奶油奶酪由酸牛奶制成
2. 这家工厂每天处理 45 万升牛奶
3. 牛奶冷却分离成脱脂奶和奶油

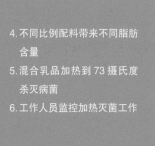

4. 不同比例配料带来不同脂肪含量
5. 混合乳品加热到 73 摄氏度杀灭病菌
6. 工作人员监控加热灭菌工作

7. 醒目的工作流程表
8. 加进乳酸的牛奶在分离器
   中除去乳清
9. 除去多余水分的凝乳再次
   加温灭菌

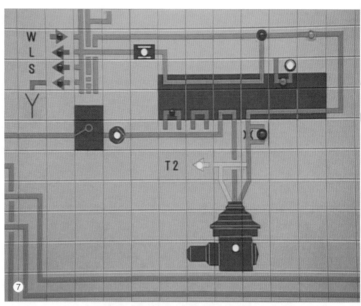

10. 加进干洋葱和干香葱等口味
11. 用氮气搅拌奶酪使之蓬松
12. 成品注入盒子后密封出厂

Cream Cheese

Would you spread soured milk on your bread? You might not think so but that's actually what cream cheese is. It's not as popular as cheddar in the UK but it's catching on fast.

At this German factory they process 450,000 litres of milk a day but before it can be turned into cream cheese it's got to be cooled down. Next it's separated into skimmed milk and cream these are then blended back together in varying quantities to give cream cheeses with a range of different fat content.

The dairy mix needs to be pasteurized. It's heated for twenty seconds at 73 degrees to kill off any germs. Then lactic acid is added. When the mixture is heated the lactic acid spreads through the milk making it thicker and turning it sour.

It's then poured into a separator to remove the excess liquid which is called whey. If you remember Little Miss Muffet you might be able to guess that what's left behind is the curd.

This is what the cheese will be made out of but it needs a little salt. Or a little bucket of salt and then it's heated again just in case any germs made it through.

When you think of cream cheese, smoke salmon and a bagel might come to mind but this factory caters for a wide range of taste. This cream cheese is going to be mixed with dried onions and chives. If you've got a date in the evening you might want to think twice about having this for lunch.

Cream cheese is different from other cheeses it's not allowed time to mature as it's meant to be eaten fresh.

Despite its popularity in Europe it was actually invented in the United States. In 1872 a New York farmer developed a richer cheese than ever before. And this became known as cream cheese. The final step is to whip up the cheese with some nitrogen gas this helps to increase its volume and make it easier to spread.

The tubes are made out of the plastic on these two rolls. It's heated and then a machine punches it into shape. Next each pot gets a perfect portion. That includes the weight watcher variety. They're sealed, chilled for twelve hours and then ready to hit the supermarkets.

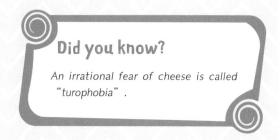

Did you know?

An irrational fear of cheese is called "turophobia".

卡芒贝尔奶酪

扫描二维码，观看中文视频。

欧洲各地都挤牛奶，但这种牛奶不会浇在玉米片上，它有更高贵的使命，在短短一个多星期内变成卡芒贝尔奶酪。每天超过 10 万升的鲜牛奶被送到这家工厂。牛奶运送到罐子里，在那里被加热以杀死所有细菌。

添进乳酸让牛奶变酸。然后掺入霉菌孢子。虽然它们只有千分之一毫米大小，但仍然对最终产品有很大影响。变酸的牛奶被倒入 2000 升容积的大盆里再加入酶，使混合物变黏稠。混合物被静置，分离成凝乳和乳清。

首先用正方形网格纵向切割。接着另一个网格沿横向切割。凝乳立方体沉到底部，液态乳清留在顶部。用机器把所有的东西过筛，将大部分乳清排掉。凝乳块被紧紧压到圆锅里，它们看起来开始像奶酪了。每个锅都要翻转几次，除去剩余的乳清，进一步压缩凝乳块。这些锅被叠放在托盘上再被翻转几次，加入其他 20 万个锅的行列，然后静置 10 小时。

下一步是盐浴。这有助于奶酪储存。经过清洗和放置，现在的奶酪已经足够坚固，能够从模具中取出来。该让先前加进的孢子开始活动了，奶酪被送去酝酿和成熟。因为暴露在氧气里，孢子沿奶酪的表面生长。卡芒贝尔奶酪被定期翻转，以避免霉菌生长到格栏上。

霉菌的作用是减少酸在奶酪表皮上的积累，并使奶酪内部成熟，产生一种奶油般的质地。使得这款奶酪具有独特的滋味和香气。

经过 7 天，奶酪准备进行包装了。你可能习惯于看到楔形或者整块圆形的卡芒贝尔奶酪，但这里被切成两半独立封裹。两块半圆形奶酪团聚在一个盒子里，准备食用了。能和饼干一起吃，夹在三明治里烤，甚至在奶酪火锅里熔化。

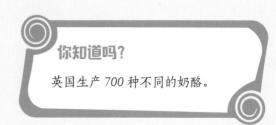

你知道吗？

英国生产 700 种不同的奶酪。

1. 工厂每天需要 10 万升鲜牛奶
2. 牛奶在罐中加热杀菌
3. 变酸的牛奶加入酶后变稠
4. 用网格纵向切割
5. 另一组网格沿横向切割
6. 留下凝乳块沉到底部

7. 凝乳块被压到锅形模具里

8. 模具叠放在托盘上多次翻转

9. 静置 10 小时后进行盐浴

10. 奶酪从模具中取出熟化

11. 经过 7 天奶酪已经成熟

12. 成品奶酪包装出售

Camembert Cheese

Cows are milked all over Europe, but this milk won't be poured over corn flakes-it's got a higher calling and in just over a week it's going to be Camembert cheese. Every day over 100,000 litres of fresh milk are delivered to this factory. The milk is transported into tanks where it is heated to kill any bacteria.

Lactic acid is added to sour the milk. Then mould spores are added. Although they're only a thousandth of a millimetre in size they'll still have a big effect on the final product. The soured milk is poured into large tubs that hold up to 2000 litres and an enzyme is added which will thicken the mixture. The mixture is set and can be separated into curds and whey.

First a grid of small squares cuts lengthwise. Then another grid cuts sideways. The cubes sink to the bottom and the liquid whey deposits on top. It's all sieved through a machine and most of the whey is drained off. The cubes are then firmly pressed into round pots and for the first time it starts to look like cheese. Each pot is flipped several times this removes any remaining whey and compresses the cubes further. Stacked on trays the pots are turned several more times before they join 200 thousand other pots for a 10 hour rest.

The next step is a salt bath. This helps to preserve the cheeses. After a wash and a rest the cheeses are now solid enough to be removed from their moulds. It's also time for the mould spores that were added earlier to kick into action as they're sent off to mature. As they're exposed to oxygen the spores grow on the surface. The camembert is regularly turned to avoid the mould growing to the grid. The effect of the mould is to reduce the acid growing on the cheeses skin and to ripen the inside giving it a creamy texture. This gives the cheese it's distinctive flavour and smell.

After 7 days the cheese is ready to be packed. You might be used to seeing Camembert in wedges or full round blocks but here it's cut in halves and they're wrapped individually. The halves are reunited in a cardboard box and then they are ready to be eaten off crackers, toasted in sandwiches or even melted down at a fondue party.

Did you know?

There are 700 different cheese produced in Britain.

帕马森奶酪

一般情况下牛会被打上烙印，但这是来自意大利的帕尔米加诺奶酪，通常称为帕马森奶酪。帕马森奶酪的生产从牛开始，但我们只需要牛的奶。意大利曼托瓦的这家乳品厂每生产一块奶酪需要半吨牛奶。奶酪在铜质的大桶里做成。每个桶能盛990升牛奶——足以制作两个巨大的帕马森奶酪圆盘。这些奶牛需要吃很多草才能跟上生产节奏。

帕马森奶酪是一种硬奶酪，所以需要牛奶凝固，这个过程最理想的温度是33摄氏度。为了让牛奶凝固，制作者会使用凝乳酶。这种酶来自牛犊胃里，将它倒入牛奶放一段时间便会发挥魔力。由于欧盟法律对商标的严格规定，帕马森奶酪只能在特定的意大利地区制作。这意味着每个厂家必须尽可能多的生产奶酪满足全世界的巨大需求。

随着凝乳酶生效，领班师傅发现牛奶的黏稠度有了微妙变化。工人保持密切关注，他们知道何时开始下一个步骤。当时机合适，他们就开始工作。用定制的刀具切割酸奶状的产物，将它分成大块。以便奶酪凝乳从乳清中分离。3分钟之后，升高温度，凝乳和乳清分开。固态的凝乳离开上部的液态乳清，沉降到底部。

这个定制的切刀价值4000多英镑，它设计的功能是把桶底的大奶酪切割成两半。在帕马森奶酪行业里，高新技术与传统工艺并行。

此刻，奶酪师傅拿他的大木桨捞起两半奶酪，让同事们用布包裹起来。如果他们把奶酪留在桶底，就需要有人要爬进去把它取出来。相比之下这种方法容易得多。工人们现在可以吸出所有的

旧乳清，为大桶处理下一批牛奶做好准备。

奶酪一旦取出后，就被包裹起来，顶部放上重物。这样能挤压出剩余的液体。作为一种硬奶酪，帕马森所含的液体越少越好。它将以这样的状态，在特氟隆模具里待上8小时。

当奶酪延展开来后，乳品厂的名字就印到了侧面。大约24小时后，特氟隆模具被金属模具取代。奶酪在其中沉降并变成有特点的轮盘造型，顶部和底部平坦，侧面带有弧度。

3天后，模具中的奶酪可以洗洗澡了——一个盐浴，此举确实能改善奶酪的气味。奶酪被留在盐水里待上1个月，然后取出干燥。这个过程能增进奶酪最终的味道。

一旦洗完了盐水澡，它们就进入熟化室。这里的设施总价值约1700万英镑，刚出浴的帕马森奶酪将加入其他奶酪的行列。奶酪圆盘将花费长达2年的时间在这里慢慢成熟。但为了避免发霉，必须至少每2周翻动一次。翻动奶酪非常枯燥和辛苦，所以要用机器人来完成。如此长时间从事这项工作，看来机器人也需要翻修清理一下了。

在成熟的过程中，工人对奶酪保持密切监视。制奶酪的领头工匠用他的标准的槌头随机

你知道吗？

在意大利商店，帕马森奶酪是最容易失窃的。被小偷盗走的东西十分之一是这种流行的意大利面佐料。

敲打奶酪。他训练有素的耳朵能分辨出好的和坏的帕马森奶酪。

他用一个小螺丝锥测试样本，以确保奶酪完全成熟。当他对奶酪的达标感到满意，便会给奶酪烧上信赖的烙印。从不起眼的步骤开始，通过幽暗的盐水浴，生产出了世界上最受欢迎的意大利面佐料奶酪。

1. 帕马森奶酪在铜质大桶中制作
2. 奶酪制作车间
3. 用特殊刀具把奶酪切开
4. 用布从桶中捞取奶酪
5. 奶酪在重物作用下渗出水分
6. 模具在奶酪上印出商标

7. 金属模具让奶酪成型
8. 奶酪进行 1 个月的盐浴
9. 奶酪放置在熟化室
10. 密切监视奶酪的熟化过程
11. 对奶酪进行最后检测
12. 用烙铁给奶酪打上烙印

Parmesan Cheese

扫描二维码, 观看英文视频。

Normally its cows that are branded but this is Parmigiano Reggiano from Italy, often referred to as Parmesan cheese. Parmesan does start with cows, but it's only their milk we want. This dairy in Mantova in Italy uses half a ton of milk for each block of cheese. It's made in these huge copper vats. Each one holds 990 litres – enough to make two giant Parmesan wheels. Those cows have a lot of grass to eat to keep up with production.

Parmesan is a hard cheese so the milk needs to be solidified. The ideal temperature for this is 33 degrees. To get the milk to solidify, the cheesemaker will use rennet. This enzyme comes from calves stomachs and it's poured in and left for awhile to work it's magic. Because of strict European laws concerning trademarks Parmigiano Reggiano can only be made in certain Italian regions. This means each producer must make as much as they can because global demand is huge.

As the rennet takes effect the head cheese maker will notice subtle changes in the milk's consistency. The workers keep a very close eye out so they know when to start the next step. When the time is right, they get to work. Using custom made cutters, they slice through the yoghurt-like substance, breaking it into lumps. This helps separate the cheese curds from the whey. After three minutes of this, the temperature is raised and the two parts separate. The solid curds fall to the bottom leaving the liquid whey at the surface.

This custom made knife costs over £4,000 and is designed to cut the big lump of cheese at the bottom of the tank in half. In the Parmesan business high-tech goes hand in hand with old school.

The cheese master now uses his big wooden paddle to lift the two cheese halves so his colleagues can wrap them in cloth. If they left the cheese at the bottom of the tank someone would have to climb inside to get them out. This way is far easier. The workers can now suck out all the old whey making the tanks ready for the next load of milk.

Once the cheese is removed, it's wrapped up and a weight is put on top. This squeezes out any excess fluid. As a hard cheese Parmesan needs as little fluid as possible. It will remain like this for 8 hours in a Teflon mould. As the cheese spreads out, this imprints the dairy's name into the sides.

After about 24 hours, the Teflon form is substituted for a metal one. Here it will sink down and take on the characteristic wheel shape with flat top and bottom and curved sides.

After three days, in their moulds, these cheeses could really do with a bath. A salt bath. This process actually improves their cheesy smell.

The cheese is left in this salty brine bath for a month before its taken out to be dried. This helps improve the cheese's final flavour.

Once its time to get out of their shared bath water, they make their way to the ripening room. The contents of this room are estimated to have a total value of 17 million pounds and our freshly bathed Parmesan wheels are about to join them. The wheels will spend up to 2 years in here maturing slowly. But to avoid growing mould, they have to be turned at least once every two weeks. Turning this many cheeses would be very dull and very hard so a robot is used instead. Although, after doing this job for such a long time, it looks like the robots could also do with some turning.

As it matures, the staff keep a close eye on the cheeses. Using his official hammer the head cheese-maker will tap on a random. His expert ear knows the sound of a good parmesan from a bad one.

He'll use a little corkscrew to test a sample to ensure the cheese is maturing nicely. When he's satisfied its up to scratch, he'll fire up his trusty brand and mark the cheese. From it's humble beginnings, via some rather dark and briny bathwater, the world's favourite pasta topping is born.

Did you know?

Parmesan is the most shoplifted item in Italy. 1 in 10 of the things taken from stores by light-fingered shoppers is a pack of this popular pasta topping.

甜甜圈

扫描二维码，观看中文视频。

候默·辛普森喜爱这种食品，配一杯热咖啡，吃起来味道好极了。甜甜圈有多种口味，但它们是怎样做成的呢？它们像这样在工厂里开始制作。第一种成分是酵母。这有助于面团发起给甜甜圈一种轻微蓬松的质感。在温暖条件下酵母的作用更快，但温度过高酵母细胞就会死亡。所以在炎热天气要把冰加进混合物中。其次是面粉，大量的面粉。这个工厂每天要做2万个甜甜圈，消耗0.75吨面粉。

你也许不知道，甜甜圈厨师还必须是业余气象预报员。如果空气太潮湿厨师应该在面粉里加较少的水。所以每次开始做新面团的时候必须检查晴雨表和气象图。如果眼看潮湿天气就要来临，他需要调整面粉混合物，否则甜甜圈就发不好。等到他对适合环境的配方感到满意，就是该做另一批甜甜圈的时候了。

这块面团足够做750个甜甜圈。厨师的工作是把面团分开，把其中的气泡挤压出来。然后面团被送进滚压机轧平准备切成甜甜圈分量的小块。需要几个来回才能恰到好处。

甜甜圈中间的孔是哪里来的，这个谜从未有过恰当的解释。最流行的说法是，有个面包师的徒弟不喜欢妈妈为他做的饼，因为中间部分常常是生的。妈妈就把坚果放到中间，儿子又把它们取出来。于是挖去了坚果的面团，就变成了中间有个洞的甜甜圈。现在用机器来做孔能节约成本，再没有人把中间面团切下来扔掉。我们将在稍后对此做更多了解。

下面甜甜圈都排在平板上并送到一个温暖的橱柜里待50分钟，酵母在那里开始起作用将面团发酵。这刚好给了工人足够的时间把中间切下的面团也做成美味小吃。这只能是个不令人满意的工作，精心把恰好120个小面团成队成行排列到托盘后另一位同事收取过来倒进油炸锅里，所有的整齐秩序全都毁掉了。

在油炸小面团的同时，老板也盯着橱柜里的发起的甜甜圈。在38摄氏度的温度下，它们发得很好。再来看中间切下的小面团。在190摄氏度的豆油里炸好变脆，取出来后涂上一层又厚又黏的糖衣作为单独的甜甜圈出售。

现在轮到甜甜圈。它们同样需要在油锅里走一遭，炸好一面后再翻到另一面充分烹炸。出锅时看上去像甜甜圈了。但吃起来还不大像。现在他们需要"顶层装修"。这个过程靠手工，通常的调料包括草莓、太妃糖、奶油冻和肉桂。有的甚至用五颜六色的糖屑。

这家工厂每天生产2万个甜甜圈，运送到全国各地。于是，作为节食者的灾祸，充满诱惑的故事到处蔓延——这些令人眼馋、高卡路里、黏糊糊的甜甜圈啊。

你知道吗？

世界上最大的果酱甜甜圈是1993年1月在纽约尤蒂卡制作的，重达1.7吨。

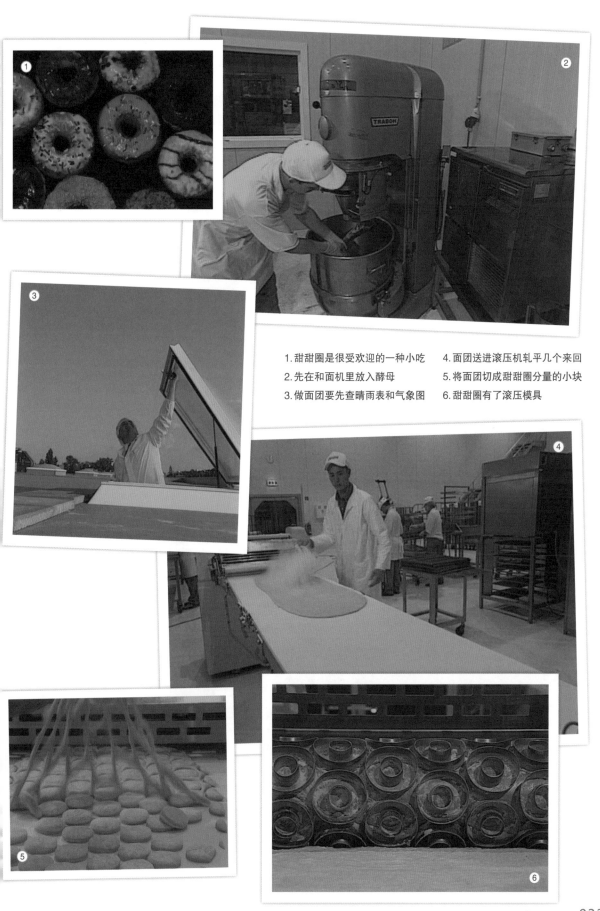

1.甜甜圈是很受欢迎的一种小吃
2.先在和面机里放入酵母
3.做面团要先查晴雨表和气象图
4.面团送进滚压机轧平几个来回
5.将面团切成甜甜圈分量的小块
6.甜甜圈有了滚压模具

7.面团切出甜甜圈的模样

8.在 38 摄氏度发酵良好

9.小面团在豆油里炸好涂上糖衣

10.甜甜圈同样要在油锅里走一遭

11.给出油锅的甜甜圈添加糖屑

12.质检员对成品进行检验

Donuts

Homer Simpson loves them and they taste great with a hot cup of coffee. Ring Donuts. There's a wide variety available, but what goes into making them? They start out in a factory like this one. The first ingredient is the yeast. This helps the dough to rise and gives the ring its light fluffy texture. The yeast works faster if conditions are warm. But if it's too warm, the yeast cells will die. So on hot days, ice is added into the mix. Next is flour and plenty of it. To make 20,000 thousand donuts, this factory uses nearly 3 quarters of a tonne of flour everyday.

Now you may not know this, but donut chefs also have to be amateur weather forecasters. If the air is too humid, the chef must use less water in his mixture so he has to check his barometer and weather maps every time he starts a new batch of dough. If it looks like wet weather's coming, he needs to adjust his mixture otherwise his donuts just won't rise properly. Once he's satisfied he's got the recipe right for the conditions, its time to turn out another batch of fresh doughnuts.

This one mix will be enough to make 750 fresh rings. This chef's job is to divide the mix and beat out any unwanted air bubbles. The dough is then sent to the rollers to flatten it ready to be cut into doughnut sized pieces. It needs a few passes to get it just right.

Now the mystery of where the traditional hole in donuts came from has never been properly explained. However, the most popular story is that a baker's apprentice didn't like the cakes his Mum cooked for him, because they were often raw in the middle. She started putting nuts in. But he also took these out…so Dough with nuts, which were removed, became donuts with a hole in the middle. A machine helps them with the holes today and to save money, no one throws the middle bits away any more either. But we'll learn more about them in a minute.

The donuts are now laid out on a board and sent to a warm cupboard for 50 minutes where the yeast can get to work and help the dough to rise. This gives the workers just enough time to make the middles into tasty snacks as well. Now this has to be an unsatisfying job. After carefully laying out exactly 120 middles onto a tray in neat rows, a colleague collects them and dumps them into a deep fat fryer where all the tidy ordering is ruined.

Whilst they're sizzling away, the boss keeps an eye on the donuts rising in their cupboard. At 38 degrees Celsius, they're growing nicely. Back to the middles. After crisping nicely in a bath of hot soya oil at 190 degrees Celsius, they're removed and coated in a thick sticky layer of a sugar water mix and sold as individual donut bites.

Now it's the turn of the donuts. They also have to take their turn in the hot oil bath. Enough to cook one side, then they're flipped over and the other side is cooked to perfection. When they emerge they look like donuts, but they don't taste of very much. So now they need a topping. This process is done by hand and popular flavors include strawberry, toffee, custard and cinnamon. Some even come with multi-coloured sprinkles.

Everyday this factory produces around 20,000 donuts to be shipped out all over the country, so that's the tempting tale behind the bane of dieters everywhere—the much-craved and calorific sticky ring donuts.

Did you know?

The biggest jam–filled donut the world was made in Utica, New York in January 1993. It weighted a massive 1.7 tons.

德国啤酒

在一个美好的夜晚，你可能会来点鸡尾酒。不过从历史上看，英国饮者往往首选啤酒。打从几千年前被美索不达米亚人发明，它直到今天一直是人们在酒吧里的最爱。特别受人喜欢的是德国啤酒，以纯度和味道著称。他们甚至有一个专门的法律叫作纯度法。只有天然产品才允许在酿造过程中使用。

用于制造啤酒的谷物是大麦。它每年由巨大的联合收割机收割，并将麦粒和秸秆分开。在啤酒厂大麦与温水混合，温度大约在 15 摄氏度，这是一种称之为萌发过程的开始。在巨型铜罐中，大麦吸收温水在种子里触发反应，这对酿造过程至关重要。大麦实际上开始生长。这种生长在麦粒中产生麦芽，是制造啤酒所必需的。经过大约 6 天大麦被干燥并翻动以防止腐烂，这时留下的看去像麦片，这就是出芽大麦。

啤酒爱好者们，啤酒的酿造过程从这里开始。发芽大麦与新鲜干净的水混合。在发芽过程中产生的麦芽现在可以和大麦自身的天然淀粉起作用。复合糖或淀粉需要被分解，这个绿色的小人是做这项工作的酶。酶与淀粉相互作用破坏了分子链而产生糖。酒精便来源于此。

2002 年英国人消费了近 150 亿品脱（1 品脱约为 0.568 升）啤酒。因此德国啤酒制造商有大量事情要做。一旦麦芽完成了它的任务并分解了所有淀粉，混合物将被送到下一个罐子里。如果本地酒家给你一杯这样的啤酒，你大概会请他换一杯新鲜的。这种混合物仍含有麦芽粉和麦壳所以下一步将清除它们。罐子的底部是一个筛子，让液体流走把麦芽留下。这些残渣会卖给农民，因为它是优良的牛饲料。

现在所有的纤维和麦壳已经被去除，清澈的液体送到另一个不同的罐子，加进下一种成分：啤酒花。啤酒花会给啤酒带来特有的苦味。啤酒花使用越多啤酒的味道就越苦。

对于啤酒来说，最后的成分可能是最重要的酵母。酵母给啤酒带来两个最明显的特性。首先是泡沫。酵母发酵释放气体使得啤酒泡沫泛起。一开始有太多的泡沫，你能得到的只是一杯泡沫。但啤酒留在罐子里在 2 摄氏度下酿造约 4 周，这使得它有时间产生香味，并由酵母发酵把糖转化为酒精。酒精是啤酒的第二个重要特征。

做老板的一大好处是，你能率先品尝每一批新酿好的啤酒——它似乎味道很好。

剩下的事就是把啤酒运到商店或酒吧，让你能来上 1 品脱。数以千计的回收瓶子运到工厂，但需要进行清洗和检查以确保安全。这台机器用灯光照射洗好的瓶子，任何有缺陷的都被送到回收箱。

这家工厂每小时能灌装 50000 瓶啤酒，随后加上新的盖子。瓶子然后通过水洗，这样新商标才能贴的牢固。最后将啤酒装箱，运载到德国及世界各地的酒店和超市。德国工业以工程的精密而闻名，对于幸运的饮酒者来说，这种注重细节的理念在啤酒生产中也不例外。

你知道吗？

平均而言，英国人每天消费近 2800 万品脱啤酒。

1.德国啤酒是酒吧里的最爱

2.德国专门有关于啤酒的法律

3.纯天然原料才能用于啤酒生产

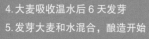

4.大麦吸收温水后 6 天发芽

5.发芽大麦和水混合，酿造开始

6.观察淀粉分解状况

7. 啤酒花使啤酒产生苦味

8. 在这里加进酵母

9. 发酵后气体使啤酒泛起泡沫

10. 酿造约 4 周便可品尝

11. 装满啤酒的瓶子进行清洗

12. 注重细节的理念处处体现

BADISCHE STAATSBR

Mindestens haltbar bis

German Beer

扫描二维码，观看英文视频。

For a classy night out you'd probably reach for the cocktails. But historically speaking the English drinker has often preferred beer. Invented by the Mesopotamians thousands of years ago, it's a firm favourite in the pubs to this day. A particular favourite is German beer; famous for its purity and taste. They even have a dedicated law called the Purity Law Only natural products are allowed in the brewing process.

The grain used to make lager is called barley. At the brewery, the barley is mixed with warm water. It's about 15 degrees Celsius and this is the beginning of a process called germination. As it sits in the enormous copper tanks, the barley absorbs the warm water which triggers a reaction in the seeds vital to the brewing process. The barley actually starts to grow. This growth spurt starts the production of malt in the grains which is needed to make beer with. After about 6 days, the barley is dried and rotated to stop it from rotting and the brewer is left with what looks like muesli. This is malted barley.

And for all you beer fans out there, this is where the brewing process starts. The malted barley is mixed with fresh clean water. Here the malt that was produced during germination can now get to work on the natural starches in the barley itself. The complex sugars or starch needs to be broken down. This green character represents the enzyme that does the job. The enzyme's interaction with the starch breaks the chain; producing sugar. This is what you make alcohol from.

In 2002 Britons consumed almost 10 and a half billion pints. so the brewers have a lot of work to do. Once the malt has done its job and broken down all the starch, the mixture is passed on to the next tank. Now, if the barman in your local served you a pint of this you'd probably ask for a fresh glass. The mixture still has all of the malted barley powder and husks in it so the next step is to get rid of it. At the bottom of this tank is a sieve and the liquid flows out leaving the malt inside. It's sold to farmers as it makes excellent feed for cattle.

Now that all of the fibre and husks have been removed, the clear liquid is sent to a different tank where the brewer adds the next ingredient. The hops. Hops will give the beer its characteristic bitter flavour. The more hops used, the more bitter the final taste.

The last ingredient is probably the most important one for beer. Yeast. This is responsible for the two most obvious characteristics of lager. Firstly bubbles. The yeast ferments which releases gas and this makes the beer frothy. At first there are too many bubbles and all you get is a glass of foam. However the beer is left to brew in the tanks at just 2 degrees Celsius for about 4 weeks. This gives it time to develop flavour and for the yeast to ferment the sugar into alcohol, the second important characteristic.

One of the perks of being the boss is that you get to sample some of each new brew that emerges. It seems to go down quite well.

All that remains is to get the beer to the shop or pubs so you can have a pint. Reusable bottles arrive at the plant by their thousands but they need to be cleaned and checked to make sure they are safe. This machine shines a light through the freshly washed bottles and any with defects are sent to the recycling bin.

The plant can fill 50,000 bottles every hour which are then capped with a fresh lid. The bottles are then passed through water so that the brand new labels will stick when they are applied. And finally they're packed into cases and loaded up to be shipped out to the pubs, bars and supermarkets all over Germany and beyond. German industry is renowned for producing precision engineering, and fortunately for those who like a tipple, that attention to detail even includes their lager.

Did you know?

On average, Britain consumes almost 28 million pints of beer every single day.

意大利面

我们都知道，最完美的意大利面，不是那些罐装的字母意面。要吃地道的真品，你必须去意大利。这家帕尔马的意面厂是全世界最大的，每天 1200 吨的硬粒小麦被送到这里。被一个巨大的真空机器从这种卡车里吸出来。其他类卡车运来的小麦粉被卸载到巨大的袋子里。

他们用的不是传统小麦。为了做出正宗的口感，他们需要用蛋白质含量高于其他小麦的硬粒小麦。将水加入小麦粉，通过搅拌和成面团。它们由电脑控制，每次都能生产出完美的意面面团。

面团从网格挤出，这些网格决定了意面的形状。圆孔网格是用来做长面的。面团从网格里被挤出 1.5 米长，然后被拉伸到 3 米。面条末端形状不整的部分被切掉，重新送回搅拌机。在拉伸和修型后，长面需要干燥以保持形状。这一步发生在 200 米长的大烤箱里。在温暖的 70 ～ 90 摄氏度环境下，长面慢慢变干。

但这家工厂并不只做长面。这张巨大的意大利面皮正被做成宽面。面落入这些碗里，然后送进烤箱。一个旋转刀斜切面管，制成通心粉。它们漏进一个振动溜槽送入筒仓，在那里干透。剧烈的摇晃有助于意面的移动，防止它们相互粘连。

当它们干燥之后，就从筒仓落下来，准备包装。看到这些巨形筒仓的体积和数量，你就大概知道有多少意面在这里生产了。长面也晾干了，但还需要被切成合适的长度，准确地说，是 26 厘米。它们按照 500 克一份进行包装。

眼前的生产规模实在让人惊叹。这家公司每年出厂 120 种不同的意大利面，总重量超过 100 万吨。

你知道吗？

2003 年，基诺·古奇制作出了有史以来最长的意大利面，长 153 米，比 3 个奥运标准游泳池还长。

1. 意大利才有地道的意大利面
2. 每天 1200 吨小麦粉送到工厂
3. 和面用的面粉是硬粒小麦粉
4. 面团经过挤压拉伸成为面条
5. 面条被拉伸至 3 米长
6. 面条送进烤箱

7. 宽面是先做出面皮再用机器切成

8. 宽面落入碗里送进烤箱

9. 容纳面条的巨大仓筒

10. 长面在容器中晾干

11. 长面被切成 26 厘米

12. 包装好的成品意大利面出厂

Pasta

扫描二维码，观看英文视频。

We all know pasta perfection can't be found in a tin of alphabeti spagetti. For the real mccoy you've got to go to Italy. This pasta factory in Parma is the largest in the World. Twelve hundred tons of durum wheat is delivered here every day. It's sucked out of these trucks by a giant accum. Wheat from other trucks is unloaded in giant sacks.

They don't just use any old type of wheat. To get the right texture they need durum wheat, which is higher in protein than other wheats. Water is added to the wheat and mixers turn it into dough. They're controlled by computers that churn out perfect pasta dough every time.

The dough's squeezed through grids which determine the pasta shape. The round holes make spaghetti. A metre and a half is squeezed through and then it's stretched to three metres. Any loose ends are lopped off and head back to the mixer. Once it's stretched and trimmed the spaghetti needs to be dried so that it holds it's shape. That happens in these whopping ovens. They're two hundred metres long. The spaghetti will gradually dry at a balmy 70 to 90 degrees Celsius.

But they don't just make spaghetti in this factory. This giant pasta sheet is being turned into tagliatelle. It get's dropped into these bowls and then heads off to an oven. A spinning blade cuts the pasta tubes at an angle to make penne. Then the Penne is funnelled along a vibrating chute towards silos where it'll dry out. The vigorous shaking helps to keep the pasta moving and stops it from sticking together.

When it's dried it tumbles out of the silos and is ready to be packed. The size and number of these giant silos helps give you some idea of just how much pasta is produced here. The spaghetti's dried out too but has still got to be cut to size , 26 centimetres to be exact. It's divided up into 500 gram portions which are then slotted into packets.

The scale of production is truly staggering. Every year this company makes 120 varieties of pasta with a total weight of over a million tons.

Did you know?

In 2003 Gino Cucci made the longest strand of spaghetti ever. At 153 metres it was longer than 3 Olympic size swimming pools.

寿司

扫描二维码，观看中文视频。

英国最近雨后春笋般冒出的寿司店，表明爱吃炸鱼薯条的英国人开始对生鱼米饭发生兴趣。寿司不仅仅只在它的故乡日本制作。在伊萨菲尔德，冰岛的一个小渔村，冷冻寿司被大规模生产。

鱼顺利抵达工厂之前，要先从海里打捞上来。这看起来就像一次普通的捕鱼之旅，但是捕到的鱼不会被熏制成腌鱼，它们会变成握寿司和卷寿司。

你可能认为他们需要罕见的蓝鳍金枪鱼或一些野生鲑鱼做寿司，但在这里他们也用寻常的黑线鳕。捕鱼线被加上重物，沉入海水下60～90米的区间，这是黑线鳕出没的深度。捕鱼线被系到一个浮标上，放在原处24小时。第二天，渔民将捕获的鱼拉上来。渔民用鱼叉将黑线鳕取出水面，然后迅速地将其杀死。离开港口9小时之后，渔民带着4吨即将成为寿司的鱼返航。寿司工厂就在港口附近，所以叉车能够将鱼直接送去加工。

按传统，寿司在日本由训练10年以上的师傅制作。但在这里很多工作都是由机器和电脑完成的。鱼头被切掉后，圆锯将鱼肉片从鱼骨上切下。鱼肉片必须是原味的，所以要放到一张明亮的桌子上接受检查，鱼肉片上任何不完美的地方，比如鱼骨碎屑都会一览无余。

虾是最受欢迎的寿司配料，当地渔民会成吨捕捞。我们都知道剥虾皮很费时间，但幸好他们研发出机器来做这项工作。机器用滚轮卡住虾壳将它剥掉而虾肉却完好无损。剥了皮的虾被铺展到传送带上更容易一目了然。它们全身每处都

要接受一台先进计算机的检查。摄像机录下传送带反射回来的蓝光，一台计算机扫描图像。根据蓝光的特性计算机可以分辨出哪是虾肉反射回来的，哪是非虾肉反射回来的。这是任何非虾肉物质的终点，它们将被一股气流吹走。还是会有细碎的虾壳漏网，只能通过人工拣出。

在冰岛，他们不仅仅准备鱼肉，而是生产完整的寿司。因此就需要大米。首先，他们淘洗大米，去除附着的淀粉。然后，把米放入一个巨形电饭锅里，煮到稍微还有点生。他们加入米醋，带给寿司米饭独特的风味。还加入糖水来平衡酸味。所有东西都在一个摇彩机般的不锈钢容器里转动，以确保每颗米粒充分入味。这些米饭看上去足够一个40人的大家庭享用，但和每年生产100吨米饭相比，这只是很小一部分。

因为寿司是生吃的，所以操作室里的卫生非常重要。三文鱼被切成适合一口吃下的大小。每片都刚好10克。他们确实把制作美味小吃的艺术变成一种精确的科学。这里，他们在制作握寿司，即鱼肉放在手指形状的米饭团上。机器将饭团压成型，并加上一滴热的日本辣根，即芥末酱。它们被整齐码放好，一个工人在每个饭团上灵巧地盖一张10克的鱼片。

这个机器用于制作卷寿司，它们是圆形，

你知道吗？

世界上最长的寿司卷是由600人在日本义井制作的，长达1000米。

外面包着海苔。机器把米饭铺在海苔片上，然后人工放上一种叫作"Tamago"的甜味煎蛋条。一些虾被绞碎再打成虾泥挤到寿司的中间。当放好食材的海苔片沿着传送带向前，它的两边会被折起来形成长长的一个整块。它目前看起来，还不像是用手指直接送入口中的小食品。但切成片后就有模有样了。

在另一头的握寿司生产线，工人们正对一些虾寿司做收尾工作。一条海苔的长带不仅好看，还能稳固米饭上面附着的鱼肉。握寿司被放到传送带上送去深度冷冻。它们将被运往欧洲各地，并且保证口感新鲜。零下 65 摄氏度深冻 12 分钟，将寿司的美味封闭其中。这家工厂甚至还做包装。他们以最快的动作装盒，确保寿司迅速回到冷冻箱。

下一次，当你蘸着酱油吃寿司时，花点时间赞叹寿司上刚好是 10 克重的生鱼片吧。品味一下经过精巧调味的日本大米饭，并想想那些筛选虾肉时被吹走和遗忘的杂碎海鲜。

1.英国人对生鱼、米饭产生兴趣
2.渔民将鱼送到寿司工厂
3.鱼片接受检查去除鱼骨碎屑
4.机器用滚轮剥掉虾壳
5.任何非虾肉物质将被吹走
6.人工捡出混在虾中的杂质

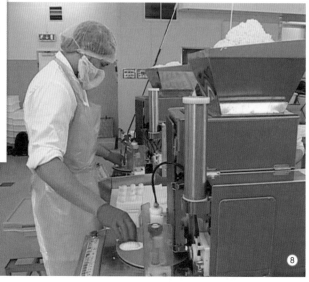

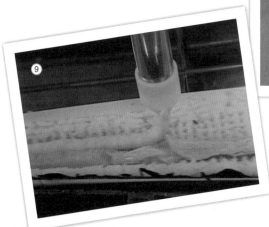

7. 在巨型电饭锅中煮米饭
8. 饭团压成型并加一滴日本辣根
9. 米饭铺在海苔上，再放煎蛋条和虾泥
10. 海苔卷起并切成小段成为寿司
11. 握寿司饭团上盖一张鱼片
12. 成品寿司送去冷冻

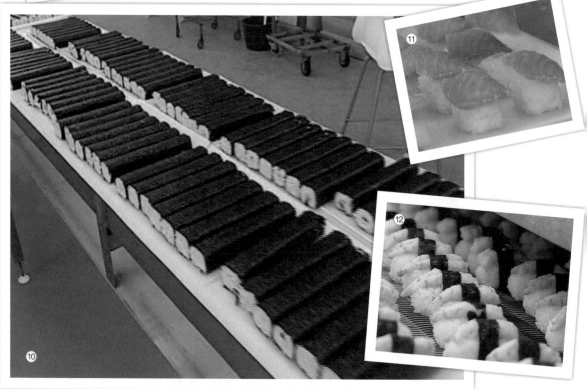

Sushi

The recent boom in Sushi restaurants has seen Britain swapping fried fish and chips for raw fish and rice. Sushi is not only made in it's traditional home of Japan. In Isafjordur, a small fishing town in Iceland they produce frozen sushi on a massive scale.

Before it gets anywhere near the factory the fish has got to be pulled out of the sea. It looks like a bog standard fishing trip but this catch isn't going to be smoked for kippers it's going to be turned into nigiri and Maki.

You might think they need the rare blue fin tuna or some wild salmon to make sushi but here they use the humble haddock too. A line is weighted to sink between 60 and 90metres. That's where the haddock hang out. It's tied to a buoy and left for 24 hours. the next day, the fishermen reel in the catch, They land the haddock with a gaffe and then quickly kill it. Nine hours after leaving the harbour they return with 4 tons of soon tobe sushi. The factory is right next to the harbour, so fork lift trucks take the catch straight in for processing.

Traditionally, sushi is made in Japan by masters who train for up to 10 years. But here a lot of the work is done by machines and computers. After the heads are being chopped off, the fillets are sliced off the bones by circular saws. The fillets have to be pristine, so workers inspect them on a brightly lit table which shows up any imperfections such as specs of bone.

Prawns are one of the most popular sushi ingredients and local fishermen catch tons of them. As we all know it takes an age to peel prawns by hand but thankfully they've developed a machine to do the job. It catches hold of the shells between rollers and removes them without touching the flesh. The shelled prawns are spread out onto a conveyor belt and they better be looking sharpish. Every spot of them is about to be checked by a state of the art computer. Cameras capture the way the blue light reflects off the conveyor belt and a computer scans the images. The computer can tell by the way the light bounces whether it is being reflected by a prawn or a non-prawn. It's curtains for any non-prawns, as they are blasted away with jets of air. Fragments of prawn shell escape and these have to be picked out by hand.

They don't just prepare the fish here in Iceland they make the finished article and for Sushi you need rice. First they wash it to remove the excess starch. Then they load it into a jumbo rice maker which cooks it till it's slightly underdone. They add rice vinegar which gives the sushi rice its distinctive flavour and sugary water to balance the acidity. It's all rotated in a sort of steel tombola to make sure that every grain soaks up the flavour. It might look enough to feed a family of forty but is only a small fraction of the 100 tons of rice they cook every year.

Since Sushi is eaten raw hygiene is at a premium in the prep room. The salmon is sliced into bite sized pieces. Each one weighing exactly 10 grams. They've really have turned the art of making a tasty snack into an exact science. Here they're making nigiri, a finger shaped pile of rice with a piece of fish on top. The machine shapes the rice and adds a drop of hot Japanese horseradish called wasabi. There stacked into neat lines and then a worker pops a 10 gram fillet on each one.

This machine is making Maki, they're the round ones with seaweed on the outside. It spreads out the rice on sheets of the seaweed and then strips of a kind of sweet omelet called Tamago are added by hand. Some shrimps are minced and then pureed before being squirted into the middle of the maki. As it passes along the conveyor belt the two edges of the maki are folded together to make a large block. It doesn't look much like finger food at the moment but after it's cut into discs it'll look the part.

Over at the nigiri line they're putting the finishing touches to some prawn sushi. A seaweed ribbon doesn't only look nice it also helps the fish stay on top of the rice.

They are popped onto a conveyor belt and heads off to the deep freeze. It's going to be shipped all over Europe and needs to arrive tasting fresh. 12 minutes at minus 65 degrees seals in the flavour. This factory even does the packaging. they box it up as quickly as possible as it can get back into the freezer.

The next time you dunk your sushi into some soya sauce take a second to marvel at the fish fillet that has been cut into an exact 10 gram portion. Savour the Japanese rice that has been delicately flavoured and spare a thought for the non-prawns that have been blown into oblivion.

Did you know?

The longest sushi roll ever was made by 600 people in Yoshii, Japan. It stretched for 1,000 metres.

威士忌

扫描二维码，观看中文视频。

没有一瓶精美的苏格兰威士忌，酒柜将是不完美的。而没有什么地方比苏格兰高地更称得上酿造苏格兰威士忌的胜地。

这是基思的斯特拉赛斯拉酒厂。酿酒的主原料是大麦，一旦收割后就被送往酒厂。第一步是生麦芽。大麦在水中浸泡 3～5 天。在这个过程中，大麦中的淀粉会被转换成糖类。此后麦芽在炉内被泥炭加热烘干，让威士忌带有烟熏香。

一旦大麦发芽，这台机器把它磨成粗粒面粉。这个巨型桶叫做糖化锅，磨出的粗粒面粉和热水在这里混合。加入酵母后，混合物开始变成酒。放进巨大的铜质蒸馏器中，被加热到 86 摄氏度。酒精挥发后，顺着这些管子向上，在那里重新冷凝成液体。这个过程大约进行半小时，酒精浓度已经达到 20%～40%。这被称作低度酒。低度酒进入第二个蒸馏器重复刚才的过程。这次蒸馏之后，就能得到浓度在 60%～70% 的烈酒了。它们被封装到木桶里变成威士忌。

在苏格兰，苏格兰威士忌制造是一个严肃的行业。依据法律，酒厂只允许把威士忌装在橡木酒桶中。新的木桶会让威士忌带上过浓的木头味，所以酒厂会使用工匠团队修整过的旧木桶。这些旧木桶可能装过波本酒，雪利酒甚至其他种类威士忌。除了能减少木桶味外，还让苏格兰威士忌带上些许其他酒的风味。

首先摘除铁箍，敲下桶盖，刮削木头。最后更换和套紧铁箍。这些木桶还要经受火的考验。工人把桶口盖住以阻断氧气供给，否则木桶会被烧成碳渣而不仅仅只是烘烤。桶的内壁形成一个木炭薄层，能使威士忌在长久熟化中增加风味和颜色。木桶被再次盖上，这次是为了把烟封进桶里面，以增加烟熏风味。

木桶用板子封上，可以准备装酒了。每个木桶能盛 250 多升。一旦桶塞被封上，威士忌将在未来很多年里不见天日。它和其他几百桶酒一起码放在黑暗的仓库里。酒必须要在橡木桶中至少熟化 3 年，才能称为苏格兰威士忌。顶级的单一麦芽苏格兰威士忌，贮藏时间要长得多。我们的酒 12 年之后才会再拿出来。在这段时间里，酒精逐步通过橡木桶蒸发，含量会从 60% 以上降低到 40% 左右。

这是一桶早年制作的威士忌。鉴定师品尝样品，以确认是否可以开桶。只要他一个点头，这些酒就能送去装瓶。经过了十多年的生产过程，装瓶厂是威士忌登上商店货架的最后一站。这样的酒厂对英国经济至关重要。苏格兰威士忌每年的出口值超过 20 亿英镑。仅从大麦、水和酵母到精美的苏格兰威士忌，体现了生活中简单的东西的确能变成最好的。

你知道吗？

"威士忌"一词来自盖尔语词组"uisge beatha"，意思是"生命之水"。

1. 苏格兰高地是酿造威士忌胜地
2. 刚收割的大麦在水中浸湿出芽
3. 麦芽在这里烘干并带上烟熏香
4. 麦芽粗粒面粉加入酵母
5. 铜质蒸馏器两次蒸馏得到烈酒
6. 苏格兰威士忌只许装在橡木桶中

7. 修整旧木桶供酒厂使用
8. 木桶经火烤使内壁形成木炭层
9. 木桶用板子封上准备装酒
10. 每个木桶能盛 250 多升酒
11. 酒在桶中熟化至少 3 年
12. 通过鉴定师评定后就可装瓶

Whiskey

No drinks cabinet would be complete without a fine bottle of Scotch Whiskey. Where better to see it being made then the Scottish Highlands.

This is the Strathisla distillery in Keith. The main ingredient is Barley. It's harvested and then taken to the distillery. The first step is malting. The barley is soaked in water for three to five days, during this time starches turn into sugars. Then it's dried over burning peat and the smoke adds flavour to the whiskey.

Once the Barley's been malted this machine grinds it down to coarse flour. This huge vat is called the mash tun. It's used to mix the ground flour with hot water. Some yeast is added and the mix starts to turn into alcohol. It's transferred to these giant copper distillery kettles and heated to 86 degree Celsius. The alcohol evaporates and travels up these pipes where it cools down and becomes a liquid again. After around half an hour of this process the fluid coming out has reached between twenty and forty percent alcohol. This is called low wine. The low wine goes through the process again in a second kettle and what comes out then is seriously strong. Between 60 and 70 percent alcohol. This will be barrelled and turned into whiskey.

Scotch is a serious business in Scotland. By law Distilleries can only store whiskey in barrels made of oak. New barrels would give the whiskey to much of a woody flavour so they recycle old ones which are fixed up by a team of coopers. These old barrels could have previously been used for bourbon, sherry or even another whiskey. As well as providing a less woody flavour they also impart a subtle character from their previous contents.

First the iron hoops are removed. The tops are knocked out, The wood is shaved. And finally the hoops are tightened and replaced. Then the barrels suffer trial by fire. The worker pops a cover on to cut off the supply of oxygen otherwise the barrel will burn to a cinder instead of just getting toasted. A thin layer of charcoal has formed on the insides which will add flavour and colour to the whiskey as it slowly matures. The cask is covered again, this time to seal in the smoke and enhance the flavour.

The barrel is boarded up and is ready to be filled. Each barrel holds over 250 litres. Once the stopper is bashed into place the whiskey won't see the light of day for many years. It's stacked among hundreds of other barrels in a darkened warehouse To be classified as Scotch the whiskey has to mature for at least three years but top single malts are left for much longer. Ours won't come out for another twelve years. Over that time alcohol gradually evaporates through the oak barrel reducing the alcohol content from over sixty percent to around forty.

Here's one they made earlier. The master taster tries a sample to check it's ready. Just a nod from him and it's off for bottling. After over a decade in the making the bottle plant is the last stage before the whiskey can hit the shelves. Distilleries like this one are vital to the UKs economy. The annual export of Scotch is worth over two billion pounds. From just barley water and yeast to fine Scotch whiskey it goes to show the simple things in life really can be the best.

Did you know?

The word "whiskey" comes from the Gaelic phrase "uisge beatha" which means "water of life".

羊角面包

扫描二维码，观看中文视频。

早餐羊角面包——英国人喜欢这种著名的法国食品，就像鸭子喜欢水。把它们配合到一起，本身几乎是一种艺术，这款受欢迎的糕点是拿什么做出来的呢？

主要成分之一是面粉。这家工厂每天使用将近 25 吨。还有黄油，每天使用 12 吨。这里不用低脂黄油，只用全脂黄油做出新鲜的羊角面包。

在这些巨大的工业搅拌机中，制作过程开始。每天在其中和出的面团，能做 130 万个羊角面包。对于大批量的面团，面包师将使用液体酵母，比固体粉状的品种更好用些。

所有混合成分最后在这个机器里糅合，加进酵母能使面团在稍后的烘烤过程中发起。各种成分完全混和后，下一台机器把面团切开。每个面团大约 25 千克重。

当面团即将准备好时，该加黄油了。但并不直接混到面团里，这是经典羊角面包制作手艺的真正秘密。生产线上的第一台机器辊压出面团。机器中送出一大张面皮，并沿着传送带通过。同时另一台机器辊压出黄油片。将它放置在面片上而不是混合到其中。面片和黄油一起小心折叠，做成类似面团黄油三明治的样子。再辊压到一起封闭起来。

现在是经典的法式点缀。首先将混合作料撒上面粉。然后辊压，并多次折叠。这个过程产生出令人垂涎的传统质感。混合物现在变成多层面团和黄油交替的超薄片。再辊压一次，并切成条状。

下个辊子实际上就是一系列的切刀，把黄油面片切成三角形。这些三角形被转成同一方向准备卷起。三角形进入机器，一个装置抓住每个面片把它卷成传统羊角面包的形状。这个自动化流程和人手所做的工作相同，只不过更快，但生产出来的东西还不够地道。这些卷起的糕点是直的，但每个人都知道，羊角面包应该像月亮一样弯曲。最后就需要人工帮忙了。他们的任务是把糕点做成熟悉的形状，用手工完成。

为了让羊角面包具有传统的金黄色泽，烘焙面团前，为它快速涂一层油。接着将它们送到冷库。

冷冻状态使它们能存储 12 个月以上。但因为饥饿的嘴巴在等待早餐，还有对这种美味的巨大需求，烤箱是这些糕点的最终去处。在 180 摄氏度，只需 20 分钟，就能把一盘艺术的黄油面团变成羊角面包，用来在法国、英国和世界各地蘸着热巧克力食用。这就是备受欢迎的全黄油羊角面包。

你知道吗？

羊角面包不用搭配黄油和果酱，热量就高达 340 卡路里（1 卡 =4.19 焦耳），这对减肥者来说简直是灾难。一个快餐芝士汉堡只有 300 卡。

1. 面粉运到工厂，这是制作羊角面包的主要成分
2. 加进液体酵母，工业搅拌机和出很大块的面团
3. 传送带上的全脂黄油被送到一台机器辊压
4. 机器辊压出面团
5. 机器辊压出黄油片放置在面片上
6. 面片和黄油片折叠后再辊压到一起并多次反复

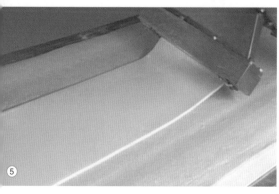

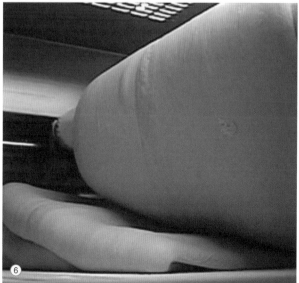

7. 黄油和面皮的超薄片
8. 一系列的切刀，把黄油面片切成三角形
9. 装置抓住面片卷成传统羊角面包的形状

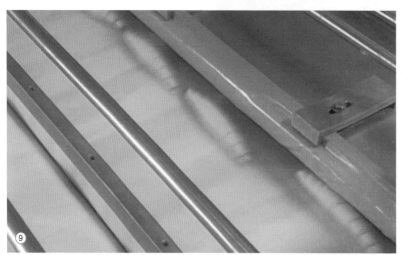

10. 工人整理好形状
11. 把它们送进烤箱
12. 在烤箱180摄氏度烤20分钟，黄油面团变成羊角面包

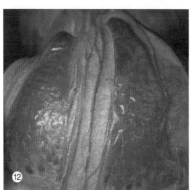

Croissants

扫描二维码，观看英文视频。

Le breakfast croissant. The Brits have taken to this famous French treat like ducks to water. Putting them together is almost an art form in itself, but what exactly goes into these popular pastries?

One of the main ingredients is flour. At this factory they use nearly 25 tons a day. And then there's the butter. 12 tons of this will be used every day as well. No diet dairy here, only the full fat variety will do to make fresh croissants.

And its here in these enormous industrial mixers that the process begins. Inside, the dough for 1.3 million croissants is mixed each day. For batches of dough this big, the bakers will use liquid yeast which is easier to use than the powdered solid variety.

A final mix of all the ingredients is being kneaded in this machine and the addition of the yeast will help the dough to rise later in the baking process. With everything nicely mixed, the next machine will carve up the dough into portions. Each one of the lumps it hacks off weighs about 25 kilos.

And with the dough nearly ready, the butter can now be added. However, it is NOT mixed directly into the dough itself and that is the real secret to crafting classic croissants. The first machine on the production line, rolls out the dough. A large sheet emerges from the machine and passes along the conveyor. Meanwhile a second machine has been rolling a sheet of butter. This is laid on top of the dough, but it isn't mixed in. Instead the dough is carefully folded over the butter making a sort of dough-butter sandwich. This is then rolled together to seal it all up.

Now for some classic French flourish. First the mixture is sprinkled with flour. It is then rolled out and layered on top of itself many times. This process creates that traditional mouth-watering texture. The mixture is now a multi-layered mass of alternate, ultra-thin sheets of dough and butter. This is then rolled out flat once more and trimmed into strips.

The next roller is actually a series of knives which cut the new buttery dough into triangles. These triangles are then rotated to face the same direction in preparation to be rolled up. As the triangles enter the machine, a matt catches each one, causing it to roll up into the traditional croissant shape. The automated process does the same job as the human hand only faster, but what emerges still isn't quite right. These raw pastry rolls are straight and everyone knows croissants are curved like the moon. But that's where a final human touch is required. Their job is to give the pastries their familiar shape, a job which is done by hand. To give the croissant its traditional golden colour, the dough is given a quick coating of oil before its baked. And next they are sent to the freezer. This freezing process means they can be stored for up to 12 months.

But with hungry mouths to feed at breakfast and huge demand for this tasty pastry, the oven is the final destination for these snacks. It takes just 20 minutes at 180 degrees Celsius to turn a tray of them from artistic pieces of buttery dough, into the croissants that will be dunked in hot chocolate all over France, Britain and the world. The popular all-butter croissant.

Did you know?

Packing a hefty 340 calories without butter or jam, the croissant is disaster for dieters. A fast foot cheeseburger contains just 300.

奶油酥饼

苏格兰，一个美丽的国家，以穿裙子的男人和奶油酥饼的故乡而闻名。酥饼是世界上最受喜爱的饼干之一。2006年这家公司在全球销售了5000万包这种酥饼。你通常能在超市或机场免税店看到它。

酥饼畅销的秘诀是什么？有人可能说是黄油。每一批面团使用近75千克的黄油，仅这家工厂每天就耗费近1吨半黄油。厨师对黄油非常重视，到达工厂的每个批次都要取样。味道必须纯正。只有最好的黄油才能使用。黄油通过口味测试后就可以添加到大桶里准备混合。就像在家庭厨房里制作饼干，第一步是把大块黄油揉在一起变成平滑细软的混合物，很容易和其他成分结合。厨师一直监视黄油直到黏稠度合适为止。然后加入盐。黄油本来含少量盐，但配方要求再多点使味道达标。盐在加进黄油之前要过筛子以确保不会有杂质意外进入混合物。

下一种配料用这样的罐子车运到工厂，其中装的是面粉，每辆约装12吨。制作传统的酥饼需要大量面粉。管道连接起来并锁定到位，但锤子是干什么的？原来面粉经常会粘到罐壁上。用锤子敲几下能让装载物松动。

回到生产厂房，秤出制作下一批酥饼的面粉，泵进搅拌器，和盐、黄油及40千克糖混合。当一切都充分混合后，将它们装进手推车并发送到炉灶旁。这项操作必须快速完成，如果面团放置太久，就会变得像石头一样坚硬而无法使用。

首先将面团碾压并切开，做好进入模具的准备。面团经过传送带顶部时会破裂和断开。易碎易折的面片比实心的面片对模具更合适。

酥饼有多种形状和尺寸，但这些传统模具生产的酥饼一直最受喜爱。每个模具做出20克一块，和那些100年前最早生产的酥饼同样大小。但酥饼只是在这些传统模具中成形。烘烤时使用较大的模具，因为混合物在烤箱中会膨胀。

现在它们已经准备烘烤了。在这个60米长的烘箱里，175摄氏度的温度把酥饼烘烤到完美状态。从另一端出来的酥饼，不论蘸茶还是单吃，都是令人垂涎的美味。它们如此受欢迎，以至在全世界许多国家都能买到。烘烤后将模具取走用来制作更多酥饼，而新鲜的酥饼送去冷却，为顾客包装到盒子和罐子里。

每一份都用塑料纸包裹保持新鲜，接着通过一个金属探测器。这是为了确保生产过程中没有工厂机器的碎片落进面团里。有问题的包裹立即从生产线上拿掉。只有完美的酥饼才能离开工厂。

最后单独的酥饼袋送去打包。并装箱分发到英国和世界各地。这就是传统的苏格兰下午茶美味——酥饼的有趣故事。

你知道吗？

"捷足"是苏格兰的传统。如果新年到你家的第一位访客带来威士忌和酥饼，那将是很幸运的。

1.苏格兰是奶油酥饼的故乡
2.使用的黄油每次都要取样检查
3.黄油投进大桶里搅拌
4.添加的糖和盐的分量要准确
5.面粉在搅拌器中和黄油混合
6.面团被碾压延展

7. 将面团压进模具中

8. 传统模具做出的酥饼每块 20 克

9. 烘烤中面团膨胀需要较大模具

10. 175 摄氏度烘烤酥饼最完美

11. 酥饼经过 60 米长烘箱

12. 通过质量检查后包装出厂

Shortbread

扫描二维码，观看英文视频。

Scotland. A beautiful country- famous for men in skirts, and the home of one of the world's favourite biscuits, shortbread. In 2006 this company sold 50 million packets of the stuff worldwide. and you can usually find it in a supermarket or a duty free shop in an airport.

And the secret behind shortbread's popularity? Well, some might say it's the butter. Each batch of dough uses nearly 75 kilos of the stuff and this factory alone uses nearly 1 and a half tons every day. The chefs take their butter very seriously and take samples of every batch that arrives at the plant. It has to taste just right. Only the best butter will do. When it's passed the taste test, it can be added to the vats ready for mixing. Just like making biscuits in a domestic kitchen, the first step is to cream the big butter blocks together, to make a smooth soft mixture that will easily combine with the other ingredients. The chef keeps an eye on the butter until the consistency is just right. When it's ready he'll add salt. The butter is already slightly salted but the recipe calls for a little more to get the flavour spot on. Before it's added to the butter it's sifted to make sure that no impurities get in to the mix by accident.

The next ingredient arrives at the plant in tankers like this one. It contains flour. About 12 tons of it in each load and to make traditional shortbread, you need plenty of it. The pipes are hooked up and locked into place, but what's the hammer for? Well the flour tends to stick to the walls of the tanker. A few blows from the hammer helps dislodge all of the load.

Back inside the production plant, the flour is weighed out for the next batch of shortbread and pumped into the mixer with the salt, butter and 40 kilograms of sugar. And when it's all properly mixed, it's loaded into a trolley and sent on to the cookers. This has to be done quickly because if the dough is left for too long, it'll go rock hard and unusable.

First it's rolled out and cut. This prepares it so it'll fit into the moulds. The dough cracks and breaks up as it passes over the peak of the conveyor belt. A crumbly, more pliable dough fits the moulds better than a solid flat sheet would.

Shortbread comes in many shapes and sizes, but biscuits produced in these traditional moulds remain firm favourites. Each mould forms a 20gram bar, the same sized biscuits as those originally produced over 100 years ago. But the shortbread is only shaped in these traditional form moulds. Larger moulds are used to cook the dough pieces because the mixture expands in the oven.

Now they're ready for some heat. Inside this 60 meter long oven, the bars are baked to perfection at 175 degrees Celsius. What emerges at the other end is irresistibly tasty when dunked in a cup of tea, or just eaten by itself. It's such a popular biscuit that it's available in many countries worldwide. After baking the forms are removed and sent back to shape more biscuits whilst the fresh shortbread is sent on to cool down and be packed into boxes and tins for the customer.

Each portion is wrapped in plastic to keep it fresh and then passed through a metal detector. This is to ensure no debris from the factory machinery has fallen into the dough during production. Any package that isn't right is immediately removed from the production line. Only perfect shortbread should leave the factory floor.

Finally individual packets are passed on for final packaging, to be boxed up for distribution across Britain and the rest of the world. That's the tasty tale of the traditional Scottish tea time treat.

Did you know?

First–Footing is a Scottish tradition. It is believed to be luckly if the first visitor to your home in New Year brings whisky and shortbread.

速食汤料

一碗暖暖的面条汤。传统的汤含有健康的蔬菜、药草，并且很有味道，但它有一个缺点。切菜、择菜和烹饪都需要时间。这可能解释了市场上的速食汤为什么会受欢迎。

速食汤的多种原料从装运卡车开始。不管你相信与否，这辆车装满了糖。生产商在许多食谱中都用到糖，因为它能改善味道。这些工人正在分拣另一种重要成分——干韭菜。去除烤焦的，其余被添加到最后的混合物里。

这里是做肉丸的地方。牛所有的部位，从腰腿肉到喉咙都被选用。第一站是绞肉机，绞到一定的黏稠度后把它送进工业搅拌机。再加入其他主要成分：干蛋白粉、面包屑和香料。这些原料会粘在一起用来制作汤里的肉丸。混合物通过这个圆筒进入机器。从机器底部出来的小球被撒上面粉并摇动以防止它们粘到一起。数百个装着鲜肉丸的架子被放入一个大烘箱，它们在其中大约停留 2 小时。

所有食材准备齐全，该把它们按照正确配方混合起来，做出不同品种的汤了。这里加入干韭菜。这个装料的小推车就是一个轮子上的秤，为工人称出每种配料的分量。和干蔬菜一起加入调味品，比如辣椒粉、盐、胡椒和大蒜粉。还有另外一种成分总被用来改善味道，这就是脂肪。脂肪开始是液体，但要加进速食汤里必须先凝固。将液体脂肪冷却到零下 14 摄氏度，这样就可以削成小薄片。如果厨师把液体脂肪倒入混合器就不会均匀分布。所有的混合物被送进这些大搅拌机，各种成分在其中充分混合。

速食汤料的生产，在英国是个庞大的产业。大约有 10 个公司生产，行业价值每年超过 73 亿英镑。竞争是激烈的，领先的制造商聘请口味品尝师，使他们的厚利企业保持鼎盛。这些品尝师的反馈意见将决定哪些新口味可以生产并登上超市货架。

回到生产线上，汤料准备包装了。但肉丸呢？这台机器处理肉丸。它们被向下送到料斗中，然后被定量分配到每份包装中。这些料斗测出合适的分量，装足后就释放出来。肉丸落下时，汤粉和面条也一同加入下面的料包中。这部分生产线的最后一个机器用热夹子把料包顶部密封起来，汤料到此完成。

每个料包放入盛有相同汤料的盒子里，它们被运到工厂周围的储存设施。从鸡肉面条汤到蔬菜通心粉汤，市场上汤料的味道大不相同，工人要确保不同品种分别装箱。一家公司声称，它生产的一种著名速食汤，每年被吃掉 2.5 亿多杯。

这是个集约化的工业流程，让消费者把一包汤料粉变成一杯热气腾腾的汤。

你知道吗？

在挪威探险家罗尔德·阿蒙森 1911 年前往南极的船上，有一位特殊的乘客——一包速溶汤。

1. 速食汤既营养又美味
2. 糖是必不可少的配料
3. 干韭菜是汤的重要成分
4. 牛的所有部位都能搅成肉馅
5. 肉馅加入蛋白粉、面包屑和香料
6. 机器里出来的小肉丸被撒上面粉

7.放着肉丸的架子被送进烘箱

8.按不同配方做出不同品种汤料

9.低温下将凝固脂肪切成薄片

10.脂肪和所有配料充分混合

11.品尝师决定何种汤料投入生产

12.肉丸和汤粉面条装进料包

Instant Soup

扫描二维码，观看英文视频。

A warming bowl of noodle soup. The traditional soup is full of healthy vegetables, herbs and lots of flavour, but there's one drawback. All that chopping and reducing and cooking do take time. But this may help to explain the popularity of all the instant soups available on the market today.

Life for many of the ingredients in instant soup starts out in trucks and believe it or not, this one is full of sugar. Soup producers use some sugar in many of their recipes because it adds flavour to the final taste. These workers are sorting another important ingredient. Dried leeks. Any burnt ones are removed but the rest will be added to the final mixture.

And this is where the meatballs are made. All different parts of the cow are used from the haunch to the throat. Its first stop is the mincer. Once it's all ground down to a basic consistency, it's added into an industrial mixer. Other key ingredients are added; dried egg-white powder, breadcrumbs and spices. Together they help bind everything to make the meatballs for the soup. The whole mixture is then passed down through this cylinder and into the meatball machine. The little balls that emerge from the bottom are sprinkled with flour and shaken so they don't all stick together. Hundreds of racks of fresh meatballs are then placed into a large drying oven where they will spend about two hours.

With all the ingredients prepared, now it's time to mix them into the right combinations for the various soup varieties. Here the dried leek is being added. The cart it's fed into is a large set of scales on wheels. This helps the worker to measure the right amount of each ingredient. As well as dried vegetables, he will also include spices such as paprika, salt and pepper and garlic powder. And there's one other ingredient that always helps enhance the flavour and that's fat. It starts out as liquid, but to get it into the soup it has to be solidified. The liquid fat is cooled to minus 14 degrees Celsius. That way it can be chipped into little flakes. If the cooks poured liquid fat into the mixing bowl it wouldn't disperse evenly. The whole mix is then sent to these big blenders where the ingredients are thoroughly combined.

Instant soup production is a massive industry in the UK. There are about 10 major companies that produce it and the industry is worth over 73 Million pounds every year. Competition is fierce, so leading soup-makers also employ taste testers to keep their profitable business on the boil. The feedback from these testers will determine which new flavours are deemed good enough to make it to the supermarket shelves.

Back on the production line the soup is ready to be packed up, but what happened to the meat balls? Well, this machine is taking care of them. They're fed down into the hoppers that will distribute a specific amount of meatballs into every packet. These hoppers measure the right weight and when there's enough inside, they're released. As they fall they're combined with the soup powder and the noodles which are being also added into the packets below. The final machine in this part of the production line seals the top of the packets with these heated grips and the soup is ready to go.

Each packet is loaded into a box containing the same flavour and these boxes are then shipped around the production plant to the storage facility. With a huge variety of flavours on the market from chicken noodle soup to minestrone, workers make sure that the different varieties are boxed separately. 1 company claims that more than 250 millions mugs of its famous instant soup are consumed every year.

And its an intensive industrial process that calls consumers to turn to a packet of powder into a steaming cup of soup.

Did you know?

An instant soup packet was found on the ship that took Norwegian explorer Roald Amundsen to the South Pole in 1911.

干法腌火腿

来认识一下黑猪。这种特别的上乘小猪是不会用来做熏肉或香肠的。相反它会成为金钱能够买到的最好美味之一——库拉特罗干腌火腿。这种昂贵佳肴的制作始于意大利北部波河河谷，一个以农业和农场而闻名的地区。

黑猪过着养尊处优的生活。事实上按普通猪的标准来看，甚至称得上豪华。如果是用来做香肠的猪，就只会留在猪圈里吃加工饲料，但这些家伙能够在甜玉米地里四处游荡并尽情饱餐。不幸的是，只要它们体重超过150千克就该屠宰了。

打从宰杀之日，干燥腌制猪肉的过程随之开始。不同于薄片火腿，库拉特罗火腿被视为珍品，所以屠宰者只选用猪肉最好的部位。干燥过程的第一步是备好生肉。制作者为肉做快速吸脂手术，并削掉多余的脂肪层。但不会切除全部的猪油，因为要留下一些有助于干燥处理。特别选用的这个部位猪肉非常娇嫩，所以要捆扎起来帮它度过漫长的干燥过程。

接下来是盐，大量的盐。腌制火腿就要把它包在厚厚的一层盐里，水分被提取出来并且改善火腿最终的味道。腌制者还使用胡椒提味。和所有的传统食谱类似，还要在肉上泼洒很多的葡萄酒。第一阶段已经完成，现在把它放在一旁，静置一个星期左右。在阴凉的不见阳光的室内，肉会慢慢变成红色。意大利美食几乎总是和红酒携手同行，在一周时间内制作者每天要用更多的酒和大蒜给腌肉揉搓一次，赋予它额外的味道。

下一阶段是加上保护性覆盖。将肉包裹在猪自己的膀胱里，这听起来很恶心，但却是传统配方，而且确实有效。然后，膀胱被紧密缝合，但又会扎得浑身是洞。如果完全气密，水分无法逸出，火腿就不能干燥腌制。最后把腌肉捆起来。不是为了防止它逃跑，而是为了使它具有库拉特罗火腿传统的梨形。

这就是制作者在这段时间里完成的劳动。现在该让潮湿的意大利空气接管工作了。

这个奢侈品火腿需要几个月时间来干燥，但因为被包裹在盐里所以并不会变质。不过熟化过程就是另一回事了。在整个意大利只有8个村子能够做库拉特罗火腿。这是因为只有这些村子具有特殊的气候，适合制作过程中所需的特殊霉菌生长。没错，就是霉菌。

当火腿在熟化间挂起来后，暴露在特殊的霉菌孢子环境中，它们生长在这些村庄的潮湿墙壁上。温暖潮湿的空气正好适合这种特殊的霉菌繁殖。它附着在外层的膀胱壁上，释放出一种酶到腌肉里。这些酶随后分解火腿内的蛋白质和脂肪，产生独特而鲜美的味道。潮湿空气和霉菌的结合，还有长时间的腌制流程，赋予库拉特罗火腿独特的味道，从来没有被人工复制出来。

你知道吗？

世界上最古老的干腌火腿可以在弗吉尼亚州的史密斯菲尔德食品公司找到。它已有105年的历史，并且生霉。但是如果除去霉菌它还可以被安全食用。

帕尔玛火腿批量生产的原理类似，但因为需求量大，所以以更大规模和更快方式进行。这就是为什么帕尔玛火腿与库拉特罗火腿相比要便宜得多，在超市随处可见。

一旦火腿腌制过程结束，农民可以从熟化室中把它取出并清理干净。发霉的火腿不是特别好卖。清理掉所有的孢子，农民可以用食指轻弹来测试腌肉的坚实度。他的专家耳朵能分辨出好火腿和没有熟透的火腿。

如果通过测试，就能拿掉网兜，打开膀胱袋子，看看里面的肉腌得怎么样了。有时霉菌已经渗透到膀胱袋子里面，农民要用手边的酒清理掉不需要的霉菌孢子。最后，可以坐下来享受自己的劳动果实了。

如果你想要试试这种世界上最贵的火腿，需要花费大约400英镑（1英镑≈8.9854元人民币），并且到非常高档的熟食店，因为你不会在本地超市找到这种食品。

1. 火腿原料是甜玉米地里的黑猪
2. 火腿选用猪肉最好的部位制作
3. 削掉多余脂肪并把精肉捆扎起来
4. 腌制用大量盐、胡椒加葡萄酒
5. 在阴凉避光的室内静置一周
6. 每天用酒和大蒜揉搓一遍

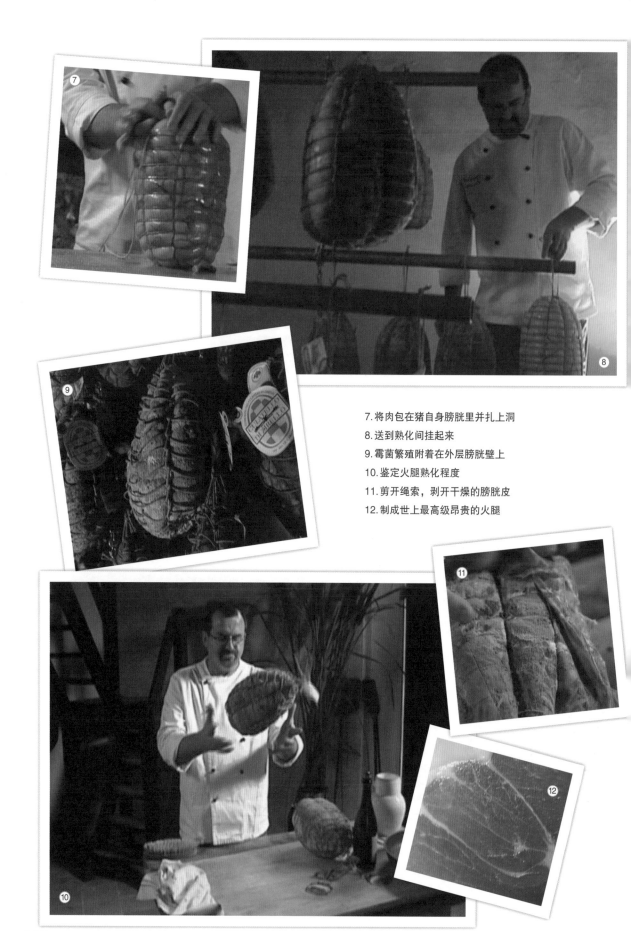

7.将肉包在猪自身膀胱里并扎上洞

8.送到熟化间挂起来

9.霉菌繁殖附着在外层膀胱壁上

10.鉴定火腿熟化程度

11.剪开绳索，剥开干燥的膀胱皮

12.制成世上最高级昂贵的火腿

Dry-cured Ham

扫描二维码，观看英文视频。

Meet the black pig. This special porker is too good to be turned into bacon and sausages. Instead he'll become one of the finest delicacies money can buy. The Culatello dry cured ham. This expensive delicacy begins life in the Po river valley in Northern Italy, an area famous for its farming and agriculture.

The black pig leads a very pampered lifestyle. In fact by normal pig standards it could be called luxurious. If he was going to become sausages he'd be stuck in a pen eating processed feed, but these guys get to roam around fields of sweet corn eating as much as they can find. Unfortunately for the pigs though, as soon as they get heavier than 150 kilos, it's time for the chop.

Once the animal has been butchered the process of dry-curing can begin. Unlike wafer thin ham, the Culatello is a considered a delicacy, so the butcher selects only the best cuts of meat. The first stage of the drying process is to prepare the meat. The butcher performs some quick liposuction and trims away a layer of excess lard. He won't remove all of it though as some is needed to help the drying process. The particular cut of meat he uses is very fragile so he has to tie it up so to help it survive the long drying process.

Next comes salt and plenty of it. To cure the ham, it's wrapped in a thick layer which draws moisture out of the meat and adds to the final taste. The butcher also uses peppercorns to enhance the flavour. And in the tradition of all good recipes he also splashes plenty of wine over the meat. The first stage is complete and he will now leave it on the side to rest for about a week. The meat will slowly turn red as it sits in a cool room out of the sun. Italian cooking and red wine seem to go hand in hand and during this week the butcher will massage the meat once a day with more wine and garlic to give it extra flavour.

The next stage is a protective covering. The meat will be wrapped in the pig's own bladder which sounds nasty, but this is a traditional recipe and it does do the job well. The bladder is then sewn up tightly, but then he fills it full of holes. If he left it airtight, none of the moisture could escape and the ham wouldn't be dry-cured. Finally, the meat is tied up. This isn't to stop it escaping, but it gives the Culatello ham its traditional pear shape.

And that is all the butcher has to do for the time being. Now it's time for the humid Italian air to take over the job

This luxury ham takes a few months to dry out, but it doesn't go off because of all the salt it's been wrapped in.

However it's a different story during the ripening process. In the whole of Italy there are only 8 villages where Culatello hams can be made. That's because these are the only villages which have the right climate to grow the right mould that the process needs. That's right, mould.

Once the ham has been hung in the ripening room, it is exposed to the particular mould spores that grow on damp walls in these villages. The warm humid air is just right for this particular mould to grow. It attaches itself to the external bladder and releases enzymes into the meat inside. These enzymes then break down proteins and fats within the ham to help give it, it's unique and delicious flavour. The combination of the humid air, the mould and the long curing process give Culatello a unique taste which has never been properly reproduced artificially.

The mass-production of Parma Ham works in a similar way, but it's done on a huge scale and much faster because of demand. This is why Parma is far cheaper and more readily available in supermarkets than the Culatello.

Once the ham has reached the end of its curing process, the farmer can remove it from the ripening room and clean it up. Mouldy ham doesn't sell particularly well. With all the spores cleaned off the butcher can test the meat's firmness by flicking it with his fore-finger. His expert ear can tell a good ham from an unripe one.

If it passes the test, the butcher will remove the string bag and open up the bladder to see how well the meat has matured. Sometimes the mould has penetrated inside the bladder, so the butcher will use a splash of whatever wine he has handy to clean off any unwanted spores. At last he can now sit back and enjoy the fruits of his labour.

And if you fancy trying one of the most expensive hams in the world, you'll need about £400 and a very up-market delicatessen, because you won't find this in your local supermarket.

Did you know?

The world's oldest dry-cured ham can be found in Smithfield, Virginia. It's 105 year old and very mouldy but would be safe to eat if the mould was removed.

能多益巧克力酱

扫描二维码，观看中文视频。

我们都知道巧克力美味可口，想出把巧克力涂在面包上吃的人是个天才。但你知道可可豆和榛子是怎样来到面包上的吗？

为了制作这种流行的坚果巧克力酱，我们首先需要可可。可可豆生长在科特迪瓦这样国家的树上，工人要在极端高温下辛劳工作收获可可豆。可可豆在热带的阳光下摊开，10天后就完全晒干了。可可豆被装进袋中并运到世界各地，像欧洲这家巧克力酱工厂。

现在可可豆被送往烤炉。可可豆有几乎50％的可可脂，焙烧使它们缩成液态糊状。然而这种可可脂在室温下会凝固，你永远无法把它涂到面包上，因此需要被去除。

为了脱掉可可脂，糊状原料被送进巨大的压力机，把每一滴可可脂挤出来。液体可可脂能在其他产品中使用。从机器中产出的东西，是巧克力爱好者心中的天堂。每块圆盘重7千克。看起来像大块的化妆品，但实际上是纯净的、压紧的可可粉。它们被送去粉碎，准备加进混合容器中。

接下来是榛子，很多的榛子。传统配方意味着每个巧克力酱罐子里大约有相当50个坚果的分量。在坚果被允许加入前，质量控制员使用特殊铡刀把样本切成两半。虫子可能钻进坚果，或者坚果会变质，这将毁掉巧克力酱。当确信坚果完好无损后，才被送去处理。

在添加进巧克力之前，坚果需要进行清洗和烘焙，所以都送往烤箱。让每瓶巧克力酱具有相同的外观和味道很重要。因此，质量差的坚果被分拣出来，并派往其他用场。计算机控制气流，把坏的坚果吹走，只留下最好的。它们和纯可可粉、糖、香草、脱脂牛奶在一个巨大的容器里混合。成为大家喜爱的柔滑软膏。工厂已经生产出大罐的巧克力酱了，但要到达你的早餐桌上，还需要广口瓶。

这些巨大的熔炉每天用回收的玻璃生产成吨玻璃容器。巧克力酱是如此受欢迎，全球巧克力酱第一品牌的销量超过所有花生酱的总和。这个炉子要加紧工作。才能满足需求。准备就绪后，熔融玻璃被送进一台看似好莱坞科幻电影中的机器里，这便是制瓶机。一团熔融态玻璃射入模具中。它们在这里被压成瓶子的形状。下一台机器做出拧盖子的螺纹，就像这样。

为确保刚生产的瓶子完美无缺，得让它们通过一组火焰。这样做能够封住玻璃中可能存在的小孔。然后迅速将它们喷水冷却，送到一个房间里，在30摄氏度下待上2天。如果你把热巧克力酱放到冷玻璃瓶中，就会出现水汽凝结。没有人希望这样。当瓶子准备好了，便从这些管道下经过并装满。剩下要做的是密封。盖子在这里被分拣出来。只有方向正确才能操作，这个巧妙的机器只选出已经面朝上的盖子。

你知道吗？

巧克力对狗和其他宠物有毒，甚至可能致命。巧克力含有可可碱，影响它们的心脏和神经系统。

其余的必须从头再来。每个盖子里有一个封口垫片，它如何被加到一瓶新巧克力酱上呢？当装满的瓶子从这台机器下方通过时，边缘上便被涂上一层胶水。加上盖子时，胶水把封口垫片粘住，工作完毕。

所以如果你是个真正的巧克力迷，放下果酱，给自己拿一瓶世界闻名的坚果巧克力酱吧。

1. 生长在科特迪瓦的可可豆
2. 可可豆在热带的阳光下晒干
3. 可可豆含 50% 的可可脂
4. 可可豆被送进压力机
5. 压力机把每一滴可可脂挤出来
6. 压紧的可可粉被送去打碎

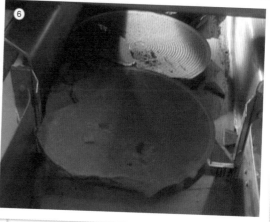

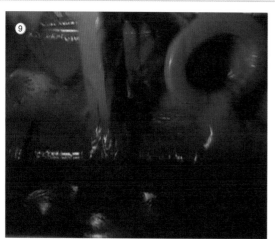

7.质检员用刀具把榛子切成两半

8.烘焙并清除劣质坚果

9.坚果、可可粉、牛奶混合成软膏

10.熔融态玻璃射入制瓶模具

11.瓶子通过火焰修整准备灌装

12.美味巧克力酱涂抹在早点上

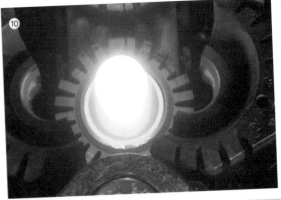

Nutella

We all know chocolate is delicious, but whoever came up with the idea of spreading chocolate onto bread was a genius. But how do you get cocoa beans and hazelnuts to stay on your toast?

To make this popular nutty chocolate spread, first we need cocoa. The raw beans grow on trees in countries like the Ivory Coast where workers have to toil in extreme temperatures to harvest them.

They are spread out to dry in the tropical sun and after 10 days they're perfect. They're then bagged up and transported round the world to places like this chocolate spread factory here in Europe.

They're now heading for the roasting ovens. The beans are almost 50% cocoa butter and roasting reduces them to a liquid paste. However, this butter solidifies at room temperature and you'd never be able to spread it onto your toast so it needs to be removed.

To do this the paste is sent to these enormous presses which squeeze every last drop of butter out. The liquid butter is sent on to be used in other products, but what emerges from the machine is a chocolate lover's idea of heaven. Each disc weighs 7 kilos. They look like big blocks of make up, but in fact they are pure, compressed cocoa. They're sent off to be crushed, ready to be added to the mixing bowl.

Next come hazelnuts, and plenty of them. The traditional recipe for a spread like this means there's the equivalent of about 50 nuts in each jar. Before they're allowed in, the quality controller uses his special guillotine to chop a sample in half. Worms can get into the nuts or they can go off, and that would ruin the spread. When he's satisfied that he's got good nuts, they're sent to be processed.

They need to be cleaned and roasted before they can join the chocolate, so they all head for the ovens. It's important that each jar of spread looks and tastes the same. So any poor quality nuts are sorted out and sent off to be used in other recipes. A computer controlled blast of air, flicks out the bad nuts leaving only the best behind. They're joined by the pure cocoa powder, sugar, vanilla and skimmed milk in an enormous tank. Here they're mixed into the smooth paste that so many people love. The factory has now got enormous tanks of spread, but to get it to your breakfast table, they also need jars.

Using recycled glass, these enormous furnaces process tons of containers everyday. Chocolate spread is so popular that worldwide sales of the brand leader outsell all brands of peanut butter put together. The furnaces have a lot of work to do to keep up with demand. Once it's ready, the molten glass is then sent to a machine which wouldn't seem out of place in a Hollywood Sci-Fi movie. This is the jar-maker. Each blob of molten glass is shot into one of these moulds. Here they're pressed into the shape of the jar. The next machine ensures that the screw thread for the lid is included and that's it.

To make sure there are no imperfections the fresh jars pass under a series of flames. This seals any small holes there might be in the glass. They're then cooled with a quick blast of water and get sent to a room where they'll sit for 2 days at 30 degrees Celsius. If you were to put hot spread into cold glass jars, you'd get condensation and moisture. No one wants that. When the jars are ready, they're passed under these pipes and filled to the brim. All that remains now is to seal them up. Here lids are being sorted. They only work if they're the right way up and this ingenious machine only uses one's that are facing the right way round already, the rest have to try again. Within each lid is an airtight seal, but how does that attach to a new jar of spread? As a full jar pass beneath this machine, it pastes a layer of glue onto the rim. When the lids are added, this glue traps the seal and the job is complete.

So, if you're a confirmed chocolate nut, put down that marmalade and grab yourself a jar of world-famous, nutty chocolate spread.

Did you know?

Chocolate is poisonous to dogs and other pets and can even be lethal. It contains theobromine which affects their hearts and nervous systems.

农产篇

稻米

扫描二维码，观看中文视频。

水稻是世界上最重要的食品之一，它构成了 20 亿人的主食。水稻最适合在蓄水的稻田里生长。这是因为它对潮湿环境的能力，比竞争对手杂草更好。

在美国加利福尼亚州，稻田在九月把水排掉，以便庞大的联合收割机可以开进来收割。收割机把作物从前面拉进去，进行脱粒。把稻谷与其他部分分离。尽管美国只生产全世界大米的 2% 左右，但由于相对较低的消费量，美国实际上是大米的主要出口国之一。

稻粒通过拖拉机转运，再由卡车送往加工厂。在收获季节，每天多达 3600 吨稻米来到这家工厂。收进的稻谷被称重并且取样用来监控每批货物的质量。一旦登记注册，稻米就被放行，开始工厂里的漫长旅程。

大功率的传送带把稻米送到 24 米高的谷仓顶部。每个谷仓可以装进 4000 吨稻米。这时大米的水分含量约为 22%，当沿着栅栏滑下去时，热空气使含水量降到约 13%，这使它变硬，足以通过下一工序而不会被损害。

稻谷通过磨坊，胶辊会把外壳除去。重量轻的糠壳被一股轻柔的气流吹走。剩下的就是保健狂热人士的选择——糙米。虽然我们大多数人喜欢不很耐嚼的白米。糙米被称为健康食品，因为棕色的糠皮中含有大量的营养物质。在这里，为了同时获得两种大米的好处，他们在高压下将米加热，驱使营养成分进入内部。然后另一组辊子可以除去褐色的糠皮，保留下营养成分。

总是会有一些讨厌的糙米会混进来，所以每粒米都经过激光扫描仪。不够白的大米被检测到，用气流快速清理掉。剩下要做的就是称重和袋装，为它前往世界各地的旅程做准备。在加利福尼亚州种植的水稻，25% 将运到亚洲，以满足那里的巨大需求。

无论糙米和白米，这种朴素无华的谷物仍将是我们星球上最重要的主食之一！

你知道吗？

有种宗教把大米视为生命的象征，并在婚礼上抛洒大米来鼓励生育。

1. 加利福尼亚州稻田把水排掉准备收割
2. 收割机对稻谷进行脱粒
3. 稻谷运到工厂称重并取样监测
4. 稻谷等级注册后准备入仓

5. 传送带把稻谷送到高大的谷仓顶
6. 装仓时热空气使稻谷降低水分含量

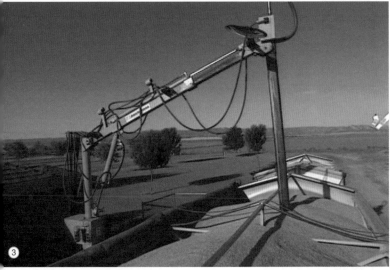

7. 磨坊胶辊除去稻谷外壳成为糙米
8. 糙米含有大量营养物质
9. 高压下加热糙米，使营养成分渗入
10. 每粒米都通过激光扫描仪
11. 不够洁白的米被气流清理掉
12. 装袋后 25% 的稻米将被运往亚洲

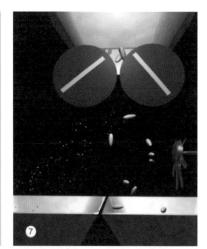

⑦

⑧

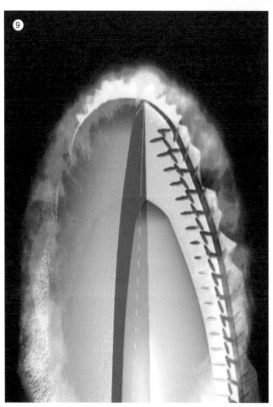

⑨

⑩

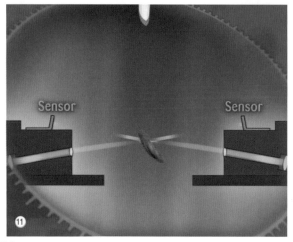

⑪

⑫

Rice

扫描二维码，观看英文视频。

Rice is one of the most important foodstuffs in the world; it forms the staple diet for over two billion people.

Rice grows best in submerged paddy fields. That's because it handles the wet conditions better than the weeds its in competition with.

Here in California, the fields are drained in September so that huge combine harvesters can move in to collect the crop. They pull the plants in the front and thresh them, this separates the grains from the rest of the plant. Despite only producing about 2% of the world's rice the US is actually one of the major exporters due to its comparatively low consumption.

The grains are transferred to tractors and onto trucks which take them to a processing plant. During harvesting season up to three thousand six hundred tons of rice arrive at this plant every day. The delivery is weighed and a sample is taken to monitor the quality of each load. Once it's been registered, the load is released to begin its long journey through the plant.

High powered conveyor belts carry the rice to the top of the 24 meter high silos. Each one can hold 4000 tons of rice. At this point, the rice's moisture content is around 22 percent as it slides down along this grid, hot air brings that down to about 13 percent which makes it hard enough to survive the next stage without getting damaged.

It's passed through a mill with rubber rollers which remove the chaff. The light chaff is then blown away by a gentle stream of air.

What's left is the choice of health fanatics-brown rice, though most us prefer the white variety which isn't as chewy. Brown rice is known as the healthy variety because the brown husk contains a lot of the nutrients. Here to get the best of both worlds they heat the rice under high pressure and this forces those nutrients inside the centre of the grain. Then another set of rollers can remove the brown husk but leave the nutrients behind.

There will always be some cheeky brown rice that makes it through, so every grain passes a laser scanner. Any not quite white enough is detected and quickly dealt with by a jet of air. All that remains is for the rice to be weighed and bagged for its journey across the world. About 25 percent of the rice grown here in California will be shipped to Asia to help meet the huge demand.

Whether it's healthy brown or healthy white this humble grain continues to be one of the most important staples on the planet!

Did you know?

Pagans saw rice as a symbol of life and introduced throwing rice at weddings to encourage fertility.

甜玉米

扫描二维码，观看中文视频。

玉米是世界上最流行的食品原料之一。我们早餐吃玉米片，晚餐有玉米棒，在电影院随意咔嚓大嚼一盒盒爆米花，破坏电影气氛。

但这里讲的是甜玉米怎样从农田来到罐头里的故事。联合收割机整日耕过这些田地，它们从地面高度把玉米秆切断，随后取下玉米棒，投进一个料斗。料斗装满后，里面的玉米被转运到卡车里，直接开进工厂加工。

卡车把小山一样的玉米倒在场院里，犁耙将它们推上传送带。在传送带的末端，玉米棒掉入一个机器里，机器内有一系列的轮子把玉米棒上的叶子剥离扯掉。玉米从传输带上流过，工人挑出烂的，让它们退出工序，用来做宠物食品。

这台机器使用光敏传感器来检查玉米棒是否都顺着相同方向。如果有的根部和尖部放反了，一个杠杆会轻轻推动它，将方向转过来。

现在是从玉米棒上剥下玉米粒的时候了。它们在推力下通过旋转刀具。刀片调节得正好适应每个玉米棒的尺寸。玉米粒被剥落下来。数以百万计的玉米粒沿传送带移动，像一条金色的河流。

在田地里待了一季，玉米需要清洗。它们泡进水里时，空心的玉米粒会漂起来，然后被丢弃。

玉米粒洗干净后，要经过筛捡。这些筛子上的孔洞只能让玉米粒通过。任何大点的东西，比如叶子碎片，都会到此止步。

这里有一个特制的摄像头，当玉米粒从前面掠过时，能发现坏的玉米粒。并用一股气流将它吹走。只有最好的玉米粒才被精选出来，准备装罐。

罐子里加进一点盐和水，来保持玉米的风味。此后，这台机器为每个罐子正好装进 140 克的玉米粒。一个摄像头用来确保每罐的分量不多不少。罐头盖子需要在真空环境里安上，才不会装进空气。这项操作在一个箱室中完成。

最后一步，是确保玉米罐头能够长时间储存。罐头进入这台机器，样子就像一个巨大的沸水锅，在 6 分钟时间里，将罐头从 25 摄氏度加热到 130 摄氏度。这个步骤为罐头杀菌，使得其中的玉米能够保鲜数年。经过 20 分钟的逐渐冷却罐头被印上保质期。最后罐头被堆放整齐并送往仓库。

200 多万听罐头在仅仅 2 个月时间里生产出来，它们被贴上标签运往欧洲各地。下次当你打开一罐甜玉米的时候，记住你将要吃到的，是通过了严格检验的最棒的玉米粒；同时想想那些只配成为狗食的劣等玉米。

你知道吗？

玉米是世界上最古老的食物之一。在墨西哥发现了超过 6000 年历史的玉米芯。

1. 收割机把玉米秆贴地面切断
2. 新鲜玉米运往车间
3. 玉米叶子被剥离去除
4. 工人拣出不合格的玉米做饲料
5. 这台机器把玉米的方向摆正
6. 旋转刀具把玉米粒剥下

7.对玉米粒进行清洗和筛选

8.玉米粒准备灌装

9.将玉米粒装进罐头

10.玉米罐在这里完成密封

11.罐头放入沸水锅加热杀菌

12.成品甜玉米准备出厂

Sweetcorn

Corn is one of the most popular food stuffs in the world. We eat corn flakes for breakfast, eat corn on the cob with our dinner and ruin films by chomping our way through boxes of popcorn at the cinema.

But this is the story of how sweet corn gets from the field to the can. Combine harvesters plough through these fields all day long. They cut the stems just above the ground and then the cobs are removed and then shot into a hopper. Once the hoppers are full the mass of corn gets transferred to a truck and goes straight off the factory so the corn can be processed.

The trucks dump mountains of corn onto the courtyard, and then ploughs push the corn onto the conveyer belts. At the end of the belt the cobs plunge down in to a machine their then pulled through by a series of wheels which grip the leaves and tear them away. As the cobs cruise past workers pick any that are rotten and for them the game is up, they'll be turned into pet food.

This machine uses a light sensor to check that the cobs are all facing in the same direction. If any are back to front a lever gives them a nudge to spin them around.

And now it's time for them to go from corn on the cob to corn off the cob. The push-through spinning knives that adjust to fit the size of each individual cob. These strip off the pieces of corn. Millions of corn niblets move along the conveyer belt like a river of gold.

After a season in the field the corn needs to be washed while their taking a dip any hollow pieces rise to the top and these will be discarded.

Now that its clean the corn gets sifted. Only bits of corn will make it through the holes. Anything bigger like bits of leaf will stay on top.

Here a special camera finds bad bits of corn as the they shoot across its path. When it finds them it blows them away with a jet of air. With only the finest corn left, it's ready to be canned.

A little salt and water is added to preserve the flavour of the corn. Then this machine loads each can with exactly 140 grams. A camera makes sure that each can has exactly the right amount. The lids need to be put on the cans in a vacuum so that no air gets sealed in. That happens in this chamber.

There's one more step to ensure that the corn will last, The cans go into this machine that is like a giant boiling pot and for 6 minutes they get heated from 25 to 130 degrees. This sterilizes the can so it will keep the corn fresh for years. Twenty minutes of gradual cooling and the best before date can be added. Finally the cans are stacked and taken to the warehouse.

Over 2 million cans are produced in just 2 months they will be labeled later and shipped all over Europe. The next time you open a tin of sweet corn just remember that you are about to dig into the very finest niblets that have made it through a rigorous inspection and spare a thought for the lowly corns which are only good enough for dog food.

Did you know?

Corn is one of the world's oldest foodstuffs. In Mexico cobs have been found that were over 6,000 years old.

玉米片

如果你以为这种流行的小食品是在墨西哥发明的，应该情有可原，但是你错了。玉米片是1940年在洛杉矶发明的，直到最近才传入英国。

不像普通薯片是用土豆做的，玉米片的基本成分是玉米粉。在这个工厂，巨大的900千克袋装玉米粉被送入混料机，与其他成分结合。包括糖、盐和淀粉。它们会花2小时，在巨大的搅拌机里揉成这样的粗面团。

不过混合物还没有准备变成人们熟悉的三角形状。大块面团不能做成薄脆的玉米片，所以首先要把面团变小。通过一系列这样的圆形刀片，将面团分割成小块为辊压做准备。

为了确保混合均匀，面团被分装进小单元隔间。这有利于任何时候都传送标准的分量。你永远不会在袋子里找到一块过大的玉米片。因为面团都是这样展开的。

面团现在被压成一个巨大的面片。面片从圆柱形切刀的下方通过，那种为人熟知的三角形就在此时出现。就像一台切饼干机，辊子的硬边划过面片，做出完美的三角。玉米片已开始成形。虽然整个流程看起来安排得很好，但总会有小面团遗留下来。但不会被浪费，碎面团被直接送到大桶里，从头再来一遍。

玉米片压好后，它们会彼此分开并被送到炉子里。但还要先经过一个小环节。

记得我们不能有太大的玉米片吗？这个工人确保玉米片不会太薄。分量过大的话，玉米片就得不到充分烹饪，太薄的玉米片会在袋子里变成碎渣。这是一件困难的工作，但必须要做好，才能得到完美的玉米片。

我们已经把玉米片放进了烤箱，但它们不会在这里待很长时间。烤架下的快速旅行让玉米片变脆并赋予它们传统的颜色，同时也意味着玉米片不用在炸锅里停留太久。

从烘箱中出来后玉米片有5分钟的时间进行冷却。这一段停顿也给检查人员提供了时间，查看有没有烤煳的玉米片。焦煳的玉米片配上调料味道不会很好。

现在该油炸了。玉米片在175摄氏度的滚油中炸上1分钟，火候正好，并且变得轻微卷曲。如果你现在开始流口水，并想抓一把尝尝，那么需要再想想，它们看起来像玉米片但却没有滋味。

市面上流行的口味如盐味和醋味道的，奶酪味和洋葱味的，咸味主导玉米片市场，制造商进行了多种口味的尝试。一种不寻常的口味叫蛤番茄，是蛤蜊和番茄的独特组合。

为玉米片调味，绝不只是把调料扔进袋子使劲摇晃。为了在每个玉米片上覆盖调味层，首先要再次涂一层油。正如这里显示的，如果没有油，所有玉米片上的调味品都将脱落。大袋的调料被加进机器顶部。从辣椒到盐和醋乃至著名的稍微不寻常的蛤番茄调料。滤出的粉

你知道吗？

世界上最大的一盘烤干酪辣味玉米片重达1.25吨，使用了616千克玉米片和280千克奶酪。

末通过管道进入机器，覆盖着油的玉米片在其中不停旋转。在这里，一个装置把调料散布在整个桶里，当玉米片旋转时，充分涂到它们表面。

英国人每年吃掉大约38000吨玉米片，足以填满骇人的193172个浴缸。

这是个很大的数量。然而当你撕开袋子，并在晚宴上招待客人之前，玉米片首先得装进袋子。这台机器对玉米片称重。根据非常精确的计时系统，每份玉米片在恰好的时刻落入袋中。

当你将一把玉米片放进美味的辛香番茄酱中时，抽空想想那位检查玉米片、为你量出精确分量的辛勤工人吧。

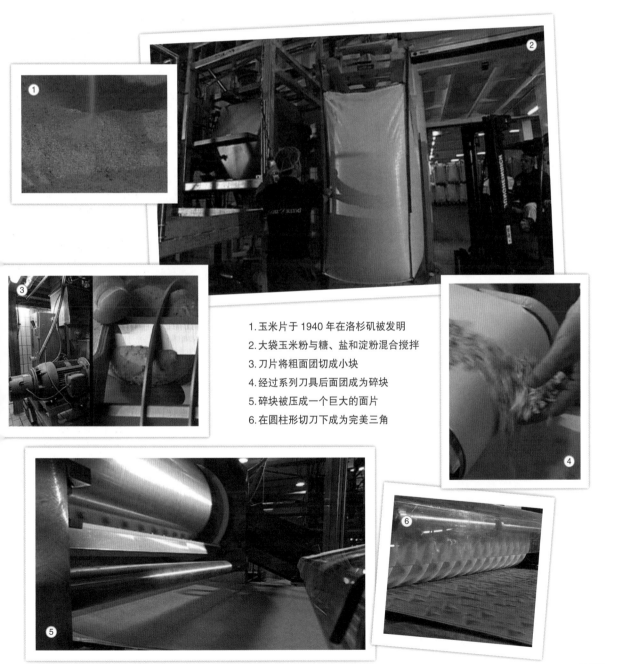

1. 玉米片于 1940 年在洛杉矶被发明
2. 大袋玉米粉与糖、盐和淀粉混合搅拌
3. 刀片将粗面团切成小块
4. 经过系列刀具后面团成为碎块
5. 碎块被压成一个巨大的面片
6. 在圆柱形切刀下成为完美三角

7.厚薄一致的玉米片被放进烤箱

8.从烘箱出来冷却后进入油锅

9.在滚油中炸过后轻微卷曲

10.覆盖调味层前再次喷一层油

11.玉米片表面涂满调料

12.对出厂产品进行检查

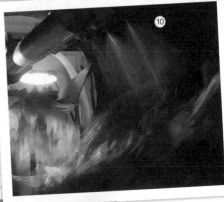

Tortilla Chips

Now you could be forgiven for thinking this popular little snack was invented in Mexico, but you'd be wrong. Tortillas were invented in Los Angeles in the 1940's but only made their way to the UK quite recently.

Unlike ordinary crisps made from potatoes, tortilla's basic ingredient is maize flour. At this factory, enormous 900 kilo bags of it are fed into a mixing machine to be combined with all the other ingredients. These include sugar, salt and starch. They'll spend a couple of hours in this giant mixer being kneaded into a thick, dough like this.

However the mixture isn't ready to be turned into the familiar triangular shape just yet. Big lumps of dough won't make crispy thin tortilla chips so first the dough needs to be reduced. It's all passed through a series of circular blades like these which breaks it into smaller pieces. This will prepare it for the rollers.

To ensure an even mixture, the dough is divided up into compartments. This helps to get a uniform quantity passing along at all times. You never get a fat tortilla chip in a bag, because the dough was spread out like this.

The dough is now rolled out into an enormous flat sheet. The sheet passes underneath a cylindrical cutter and this is where that famous triangular shape appears. Like a cookie cutter, the hard edges of the roller, slice through the dough, leaving perfect triangular piece. The Tortilla Chip is beginning to take form. Even though this process looks well organised there is always a little dough left over. But it's not wasted it's sent straight back through the dough vats to start all over again.

Once the chips have been pressed out, they are separated from each other and sent on to the ovens. But there's one small step first.

Remember we couldn't have chips that were too fat? Well this guy makes sure the chips aren't too thin. Too fat and they wouldn't cook properly, too thin and they will crumble to dust in the bag. It's a tough job, but it has to be done to get that perfect Tortilla crunch.

So we've got the chips into the oven, but they don't spend very long in here. This quick trip under the grills crisps them up and gives them their traditional colour but it also means they don't need to spend so long in the fryer.

Emerging from the oven, the chips are allowed 5 minutes to cool down. This little break also gives the inspec-tors time to check that none have been burnt. Burnt tortillas really don't go well with a good salsa.

Now it's time for the chips to be fried. They'll spend just one minute in a bath of boiling oil at 175 degrees Celsius, just enough to cook them and for the heat to curl the crisps ever so slightly. Now if your mouth is now watering and you'd love to grab a handful, think again. At this point they may look like tortillas, but they have no flavour at all.

With popular flavours such as salt and vinegar, cheese and onion and ready salted dominating the crisp market, tortilla chip manufacturers have experimented with a variety of alternatives. One unusual flavour was called clamato, a unique combination of clam and tomato.

Getting the flavour onto the chips isn't a case of just chucking it in and shaking the bag. To get a good coating onto each chip, first they're covered in oil again. As this demonstration shows, without oil, all the flavour will just slide off. Enormous sacks of the mixture are added to the top of the machine. Everything from spicy chilli, to salt and vinegar and even the famous, if slightly unusual, clamato. The powder filters down through the pipes and into the machine where the oil-covered chips are being continuously rotated. Here, a device will scatter the flavouring throughout the drum, liberally coating the chips as they spin around.

Britons eat around 38,000 tons of Tortilla chips a year, enough to fill up and amazing 193,172 bath tubs. That's a lot of chips. However, before you can rip the bag open and offer them round at your dinner party, first they've got to be put into the bag. This machine weighs up the chips. Then, following a very precise timing system, each portion is released just the right moment to fall into the bag.

So as you dunk a handful of tortillas into a tasty salsa, spare a thought for the poor guy whose been checking your chips have the perfect measurements just for you.

Did you know?

The world's largest plate of nachos weighed a massive 1.25 tons and used 616 kilos of tortilla chip and 280 kilos of cheese.

薯条

扫描二维码，观看中文视频。

在英国，我们每周要吃掉惊人的 38000 吨薯条，其中最流行和健康的品种就是烤薯条。这家位于荷兰的工厂制作薯条，它不是小作坊，而是世界上最大的薯条生产商之一。

首先土豆要按照大小分类。科研人员发现 8 厘米是理想的薯条长度。从这个网格里掉下来的土豆会因为太小而被剔除；那些个儿太大以至于不能从这个网格里掉下来的土豆，同样会遭遇淘汰的命运。在土豆世界，身材的确是重要的。

土豆皮上的大部分泥土已通过震动去掉了。剩下的泥土，会在一个装满 15000 升盐水的巨大洗涤机中清除。那些浮上水面的空心土豆，旅程到此结束。而对那些层层过关的标准土豆，现在到了蜕皮的时候。半吨土豆同时投进这个罐子。首先吹入蒸汽，接着当热土豆被倒出来时，一股冷空气让土豆皮爆裂。剩下的皮变得松散而容易除去。任何混过筛查而未被发现的腐烂土豆，会被人工检出。经过 20 分钟的处理只剩下最好的土豆。这正是他们所期待的。

当这些土豆以 60 英里（1 英里 =1.6093 千米）的时速被压过锋利的网格，便出现了一条生土豆条的河流。即使到了这一步也不能保证它们都登上烤箱托盘，仍然有一些测试需要通过。这次仍然是有关大小的检验。生薯条通过一股气流时，任何太小的将会被吹走。最终通过筛选的薯条向着炸锅进发，但它们正被电脑监视着。电脑以每秒 1000 个的速率读出图像，并在其中寻找出任何带黑点的薯条。已经通过了这么多筛选，有些倒霉的薯条仍在最后一关失败。

这些薯条终于成功通过了全部筛选，总算前程无虞了。它们在充满葵花籽油的箱体里被油炸。这些 7 米长的箱体在 1 小时内便可炸出 10 吨薯条。3 分钟后它们带着微黄色现身。火候如同从家用烤箱里取出。通过这些冷藏室后，薯条凉下来，最终在零下 20 摄氏度被冷冻。所以当你下次把完美的 8 厘米薯条蘸上番茄酱时，不妨想想它的冒险经历和它那些被抛弃的同伴们。

你知道吗？

当今英国大约有 8500 家炸鱼薯条商店，而 1950 年为 30000 家。

1. 薯条是健康而又流行的食品　　4. 土豆装入罐中充入蒸汽

2. 将过大和过小的土豆分拣出去　　5. 倒出热土豆并喷冷气使其外皮爆裂

3. 土豆进入大罐用盐水清洗　　6. 土豆压过锋利网格成为生土豆条

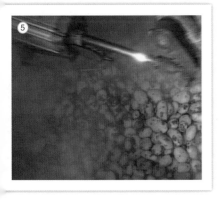

7.冲过网格后形成的土豆条河流　　10.这里每天炸 30 吨薯条

8.未达到长度的土豆条被气流吹走　　11.油锅里出来的薯条呈金黄色

9.有黑点的薯条被机器剔除　　　　　12.冷却后炸薯条进入冷冻室

Oven Chips

In Britain, we eat a staggering 38 thousand tons of chips every week and one of the most popular and healthy varieties is the oven chip. This factory in Holland makes oven chips and not just a few either it's one of the largest producers in the world.

First of all, the potatoes need to be sorted by size. The boffins recon that eight centimetres is the length of the perfect chip. Potatoes which fall through this grid are rejected for being too small and any that are too large to fall through this one suffers the same fate. In the world of the potato size really does matter.

Shaking has removed much of the soil on the potatoes skin and the rest is removed in a giant washing machine filled with fifteen thousand litres of salty water. The journey is over for any hollow spuds that float to the top. For the prime specimens that have made it this far it's time to shed their skins. Half a ton of potatoes go into this tank at a time. First steam is blown in and then as the hot potatoes are ejected a blast of cold air causes the skins to burst. What remains is loose and can easily removed. Any rotten potatoes trying to sneak through undetected are now removed by hand. After 20 minutes of processing only the finest spuds remain and here's what they've got to look forward to.

Once they have been forced through the sharp grids at speeds of up to 60 miles per hour a river of raw chips emerges. But even these have no guarantee of ever seeing an oven tray, there are still some tests left to come. It's down to size yet again. Any that are too small are blown away as they pass through a jet of air. The chips

扫描二维码，观看英文视频。

that have made the grade are on their way to the fryer, but they are being watched by a computer. It reads images at the rate of a 1,000 per second and is on the look out for any chips with black marks. After coming so far a few unlucky chips fail at the final hurdle.

At last these chips have made it, their future is secure. They're deep fried in tanks filled with sunflower oil. The seven metre tanks can fry up to 10 tons of chips in just an hour. Three minutes later they emerge slightly golden. Fried just enough to be finished off in a home oven. As they pass through these refrigerated rooms the chips are cooled down and finally frozen at minus 20 degrees Celsius. So the next time you dip you perfectly formed 8 centimetre oven chip into some ketchup just think of the adventure it has had and the ones it left behind.

Did you know?

Today there are around 8,500 fish and chip shops in Britain, compared with 30,000 in 1950.

菠萝罐头

扫描二维码，观看中文视频。

你不会设想有个整菠萝出现在比萨上，罐头菠萝块才更切实。打从发明罐头制造技术后，菠萝已经遍及全球。

在这里——夏威夷毛伊岛的西部山区，有着美国一流的菠萝生产商。这里也许看起来荒凉贫瘠，实际上肥沃的火山土壤最适于种植菠萝。

工人们在地里铺上一层塑料膜，用以保持湿度和防止虫害鼠害。新的菠萝植株，从菠萝叶冠长出。叶冠以26厘米的间隔种植，有几百行之多。工人每天需要种植9000株菠萝叶冠。这可不是像逛公园一样轻松的事情。

18个月以后，菠萝成熟了，即将准备收割。如果你只吃过菠萝罐头，也许有兴趣知道整只菠萝大约30厘米长，2千克重。和其他许多水果不同的是，菠萝一旦摘下来，就会停止成熟。但工人们知道哪些菠萝应该摘取。棕色和黄色的菠萝是可以采下的，绿色的菠萝都要再留上一天。

虽然这些菠萝大部分会被送去加工，那些卖相好的会被整只出售。送去加工的菠萝放进这台卡车里，卡车装满后开到工厂。每年有14万吨菠萝在这里加工。第一件事，菠萝被洗净和分类。这场面看起来像刺猬赛跑，其实是菠萝按个头大小决定什么时候掉下去。这样就能将它们分成不同的组。

下一步是去皮和获取果肉。只需一站式加工，菠萝就被削头、截尾、去皮、除芯。加工后的菠萝来到这里。工人们会检查果肉上是否有残留的果皮。机器把加工好的菠萝切成环状。更成熟、颜色更深的果肉味道会更甜，将被直接装入罐中，剩下的果肉会被切成块状后再装罐。

从新鲜水果到进入罐头，整个程序只需20分钟。这些菠萝被加工处理，包装完毕后，就可以用作比萨配料、水果沙拉，或者和奶酪搭配串在鸡尾酒棒上了。

你知道吗？

17世纪的时候，菠萝是作为特殊礼物敬献给皇室的。还有一幅皇家肖像画展现的就是园艺师向查理二世敬献菠萝的场景。

1. 夏威夷毛伊岛的菠萝享有盛誉

2. 肥沃的火山土壤最宜种植菠萝

3. 使用地膜保湿并防虫鼠

4. 新的菠萝植株从叶
  冠长出

5. 18 个月后到了菠
  萝采摘的时节

6. 只有棕色和黄色的
  菠萝才被摘下

7. 卖相好的菠萝会整只出售
8. 运到工厂冲洗后按个头分类
9. 菠萝处理的一站式加工

10. 工人守着机器各司其职
11. 最好的菠萝将被切成一截一截装罐
12. 菠萝罐头最后进行加热灭菌

Tinned Pineapple

扫描二维码，观看英文视频。

You wouldn't fancy one of these on your pizza, that's more like it. Since the invention of canning pineapple has been popular worldwide.

And here, in the western mountains of Maui, Hawaii, is one of America's largest pineapple producers. It might look barren but the rich volcanic soil is actually perfect for growing pineapples.

Workers are laying plastic to retain the moisture and keep the bugs and mice out. The new plants will come from pineapple crowns and they are planted 26cm apart in hundreds of rows. With nine thousand pineapple crowns to be planted, by hand, everyday. It's no stroll in the park.

After 18 months the pineapples are ripe and ready for harvesting. If you have only ever eaten pineapple out of small tins you might be interested to know the whole fruit is about 30cm long and weighs around 2 kilos. Unlike a lot of fruits pineapples stop ripening when there have been picked, but the workers know which ones to pick. Brown and yellow pineapples are good to go, any green ones are left behind for another day.

Although most of these pineapples are going to be processed, the best lookers will be sold whole. The pineapples to be processed end up in this truck. Once its full its off to the factory. One hundred and forty thousand tons of pineapples are processed here every year. First things first, they need to be washed and sorted. It might look like a hedgehog race but their size determines when they fall through the gap. That's how there get split up into different groups.

Peeling and pulping is next.

In a one stop shop the pineapples are topped and tailed. Skinned. And the cores are removed. The rest of the pineapple ends up here; the workers check the fruit for any left over skin. This machine slices them into rings. The riper, darker, and therefore sweeter pulp goes straight into a tin. The rest is cut into chunks. And also tinned.

From fresh fruit to tin the whole process takes just 20mins. The pineapples have been processed and packaged and are now ready to top pizzas, spice up fruit salads and keep lumps of cheese company on cocktail sticks.

Did you know?

In the 17th century pineapples were given as gifts to denote royal privilege. King Charles II posed for an official portrait receiving one.

开心果

扫描二维码，观看中文视频。

美味的开心果。这是个吃起来麻烦和费时的小零嘴，但如果你有耐心的话，这个工夫还是值得的。开心果富于营养和脆香可口，它们是从哪里来的？你可能不知道，世界上最大的开心果生产国是伊朗。常言道钱不会从树上长出来，但对这种诱人的经济作物来说却接近实际。在伊朗有超过 200 万人从事开心果生产，其中许多人在种植园工作。每年秋季收获，全国产量达 20 万～30 万吨。

开心果壳硬内软。在树上看起来像葡萄。开心果可以生吃，据说很可口，但不适合运输。这就是为什么我们在欧洲吃的开心果通常是干果。

把采摘的开心果运到加工厂后，需要将它们和外皮分离。这些机器能处理 80 吨坚果，但装载它们却不是一件令人兴奋的工作。首先开心果通过研磨机打破外皮。然后再通过摇动和坚果分离。接下来让开心果通过筛子，把残余的外皮除掉。

这台机器用小针去刺开心果。开心果外壳必须是张开的，以便人们容易取出内部的果仁。封闭的开心果会被拒收。有时刺针会错过坏坚果，这些女工把漏网者清查出来。剩下的便是数量巨大的生开心果。

对于伊朗国内市场，开心果都在阳光下摊开晾晒。不像英国，阳光是伊朗拥有的充足资源。需要 2 天左右把它们晒干。全球需求意味着一些生产商需要使用这种巨大的干燥罐。能将干燥进程缩短到仅仅 20 分钟。

在坚果可以离开伊朗前，必须经过彻底检查。开心果可以因真菌感染而变质，产生黄曲霉毒素——它们被确认为致癌物质。每次收获都要取样送到先进的实验室，在那里被粉碎成可以测试的浆状。稀释后的开心果汁在高科技设备中进行测试。

开心果在伊朗是巨大的产业，当局致力于不让任何东西，哪怕一丁点毒素对这项成功的生意造成任何损害。当科学家们满意后，开心果就可以发送到包装部门了。为了尽量让开心果保持新鲜，它们在踏上欧洲之旅前被真空包装。

当它们到达这里的工厂，第一步是再次仔细检查。欧洲的卫生和食品安全标准远远高于世界其他地区。小棍，石头，或任何其他碎片必须清除干净。第一站是这个太空时代的清洁器。在这台机器里，开心果都被吹起和震动。上升气流带走所有轻质杂物，同时摇晃抖动使任何多余的碎渣掉进下方的箱子。

开心果以这种形式干燥，但还是生的和没盐味的。先要洗个盐水澡：1 吨开心果和 100 升盐水混合，然后像马提尼酒那样摇动。只需要 15 分钟左右，坚果就会被盐水浸透。然后它们从摇动器里被倒出来，直接送进烤炉。坚果在这里不用花费多少时间。160 摄氏度温度下 10 分钟，就足以将它们烤熟，并产生鲜美

你知道吗？

由于开心果的含油量很高，种植者必须小心地存放开心果。如果它们变得太热，就会发生自燃。

的滋味。

开心果含有丰富的多种营养，维生素和矿物质，包括钾、镁乃至维生素 B_6，都对健康大有神益。坚果按重量秤出，落入下方的装袋机，密封后发送到商店。

从伊朗干旱平原上的种植场，来到你家附近的鸡尾酒吧，这就是有益健康和受人喜爱的盐烤开心果之旅。

1.农民在采摘开心果
2.研磨机将开心果的外皮破开
3.通过摇动把外皮分离
4.用小针刺开心果使外壳张开
5.坏果和未开口的被拣出
6.生开心果摊在阳光下晒干

7. 只用 20 分钟就让开心果干燥

8. 将开心果粉碎成浆状测试黄曲霉

9. 开心果被送往欧洲工厂

10. 在清洁器中除去杂质

11. 开心果进入烘烤炉

12. 开心果被送去包装

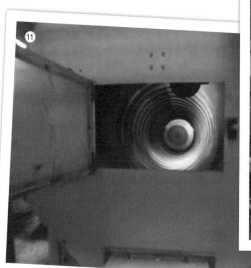

Pistachios

The delicious pistachio nut. This is a snack that can be described as fiddly and time-consuming, but if you have the patience the effort is well-worth it. They're full of energy and bursting with flavour, but where do we get them from? You may not know this, but the world's biggest pistachio producer is Iran. They say money doesn't grow on trees, but this tasty cash crop comes pretty close. Over 2 million people in Iran work in the pistachio business, many of them in the plantations. The harvest takes place in the Autumn and the country produces between 2 and 3 hundred thousand tons every year.

The pistachio grows as a soft fruit inside a hard shell. When they're on the tree they look like grapes. The nut can be eaten raw and is said to be delicious. However, they don't travel very well, which is why most of the pistachios we eat in Europe are dried.

Once the raw pistachios reach the production facility, they need to be separated from the husks. These machines can sort around 80 tons of nuts but loading them in isn't very exciting work. First the nuts are run through grinders to break up the external husk. This is then shaken off to separate out chaff from the nuts themselves. The nuts are then bumped over a sieve to knock off any determined bits of husk.

The next machine pokes the nuts with tiny needles. Pistachio shells have to open so people can get to the nut inside. Closed nuts are rejected. Sometimes, the needles miss bad nuts, but these ladies help to catch the culprits that slip through. What they're left with is enormous quantities of raw pistachios.

For the domestic Iranian market, the nuts are spread out in the sun. Unlike the UK, that is a resource Iran has plenty of. It takes about 2 days to dry them out. However global demand means some producers use these enormous drying tanks. This speeds up the process to just 20 minutes.

Now, before the nuts can leave Iran, they must be thoroughly checked. Pistachios can be spoiled by fungus which produces poisons called aflatoxins. They're believed to be carcinogenic or cancer-causing. A sample from every harvest is sent to this state of the art laboratory where they are crushed to a pulp which can then be tested. The diluted pistachio juice is tested in high tech machinery.

The pistachio industry in Iran is huge and the authorities are keen that nothing, not even a tiny toxin, does any harm to this successful business. When the scientists are satisfied, the pistachios can be sent on to the packing department. To keep the nuts as fresh as possible they are vacuum packed before they take the journey to Europe.

When they reach the factories here, the first step is to double check the nuts. Hygiene and food safety standards in Europe are far stricter than other parts of the world. Sticks stones, or any other bite-breaking debris has to be removed. The first stop is this space age cleaner. In this machine the nuts are shaken whilst being blown around. The rising air carries away any light material whilst the shaking drops any unwanted debris into a bin below.

In this form, the pistachios are dried, but still raw and unsalted. The first step is a salt bath. 1 tonne of pistachios is mixed with 100 litres of salt water and then shaken like a martini. It only takes about 15 minutes for the nuts to soak it all up. They're then poured out of the shaker and sent straight to the roasting ovens. The nuts don't spend very long in here. 10 minutes at 160 degrees is enough to roast them which brings out their delicious flavour.

Pistachios contain a rich variety of nutrients vitamins and minerals all of which are beneficial to the health, including potassium, magnesium and even vitamin B6. The nuts are divided up by weight, and dropped into the bagging machine below, sealed and sent to the stores.

So from the arid plains of the Iranian plantations to a cocktail bar near you. The healthy and popular salt-roasted pistachio nut.

Did you know?

Owing to a very high oil content, growers have to store pistachios carefully. If they get too hot, they have been known to self-combust.

香蕉

扫描二维码，观看中文视频。

据官方确认，香蕉是英国最受欢迎的水果。我们去年贪婪地吃掉了将近十亿个香蕉。但和苹果与梨子不同，香蕉是从遥远的地方进口的。在哥斯达黎加，这个种植园有 324 个足球场那么大。工人们用滑轮系统到处移动。它本来是运输水果的，但搭上滑轮能让人们省得走路。

现代香蕉没有繁育能力，因此每株新香蕉树必须从老树上切下来。这是一件非常艰苦的工作，因为香蕉树可以长到几米高。如果工人不每年补种，香蕉园将停止生产。

一旦果实开始显现，植株就必须进行修剪。底部的大紫花会争夺营养，妨碍那些肥大、美味的香蕉生长，所以工人要除掉它。留下的香蕉串，放进涂有杀虫剂的袋子里，确保它们不被吃掉。为了让昆虫不能为害，香蕉树也要喷洒杀虫剂，但不会使用飞机，而是靠更灵巧的高科技直升机，由卫星技术控制，防止当地工人自己家中也被农药喷洒。

这些香蕉的半径达 3 厘米，虽然还不成熟，却是收获的最佳时机。当香蕉长到理想的尺寸，人们就开始准备将它们送往包装厂。除去防护袋，将缓冲泡沫放置在香蕉束之间。即使水果有最轻微的擦伤，英国顾客也会把它们留在货架上，所以香蕉从一开始就受到保护。巨大的香蕉束悬挂在滑轮系统上，并加入其他香蕉长长的拥堵队列，向加工厂进发。

如果买 60 个一串的香蕉，你需要在每周购物时推一个大车。所以大香蕉串在送去清洗之前被分成小串。欧盟的法律规定了香蕉大小形状甚至曲线，工人必须确保不合标准的水果不能通过。未能达标的香蕉被扔到废品传送带上，但不会丢弃。它们用来喂养当地的牲畜，或加工成婴儿食品。

青香蕉随后送去进行另一次清洗，杀死任何试图搭顺风车，来到你家附近超市的蜘蛛。水果被贴上商标品牌，特制的箱子储存香蕉，踏上穿越大西洋的漫长旅行。这家种植园已经生产了 27 万箱香蕉。包装后由卡车运到码头。在船上它们被放进专门的冷藏集装箱前往欧洲。它们不能在甲板上享受任何阳光。这是一个 11 天的旅程，暴露在阳光下会使香蕉过早变熟。

香蕉终于被卸下，它们仍然是青绿的。这样就可以保存起来，直到商店需要的时候。香蕉被堆放在密封的成熟室里。当商店需要它们时，这些房间将充满乙烯气体。这种气体能使香蕉按照规则的、可监测的节奏变熟，这意味着商店能确切知道香蕉何时备好，何时上架。

从哥斯达黎加的异国香蕉种植园到英国的超市手推车，英国人对香蕉的确是狂热的。

你知道吗？

"猫王"以酷爱芝士汉堡而闻名，但他最喜欢的另一种小吃是炸香蕉和花生酱三明治。

1. 香蕉种植园以滑轮为运输工具
2. 补种香蕉必须从老树切下植株
3. 植株须修剪去除底部大花
4. 直升机为香蕉树喷洒杀虫剂
5. 香蕉长到半径 3 厘米可以收获
6. 收割香蕉须放置缓冲泡沫

7.香蕉束悬挂在滑轮上运向工厂

8.大香蕉串在清洗前分成小串

9.欧盟法律规定了香蕉的标准

10.青香蕉放入船上冷藏集装箱

11.乙烯能使香蕉在需要时变熟

12.成熟的香蕉摆放在超市的货架上

Bananas

扫描二维码，观看英文视频。

It's official. Bananas are the most popular fruit in the UK and we scoffed almost 1 billion of them last year. But unlike apples or pears, we import our bananas from far off places. Here in Costa Rica, this plantation is the size of 324 football pitches. The workers use a pulley system to get themselves around. It's meant to transport the fruit but catching a lift saves them from walking.

Modern bananas are sterile so each new plant must be taken from an old cutting. This is pretty tough work as the plants can grow several metres tall. If the workers didn't replant each year, the plantation would stop producing.

Once the fruit begins to emerge, the plants must be pruned. The large purple flower at the base diverts energy that would produce bigger, tastier bananas so the workers remove it. The remaining bunches are put into bags coated with insecticides to protect them from being eaten. To keep the insects at bay the crops are also sprayed but not by an airplane. They use a high tech chopper which is more agile. It's controlled by satellite technology which stops local workers being covered with pesticides in their own homes.

These bananas have a radius of 3 centimeters and although they're not ripe yet, now is the perfect time to collect them. When the fruit has reached this ideal size, these guys have the job of preparing them for the packing plant. Their protective bags are removed and foam inserts are placed between the bunches. If the fruit have even the tiniest bruises UK customers leave them on the shelf, so they're protected right from the start. Enormous bunches are loaded onto the pulley system and join a long traffic jam of other fruit heading for the processing plant.

You'd need an enormous trolley to buy a bunch of 60 bananas with your weekly shopping so now they are separated into smaller ones before being sent for a bath. Tough EU laws legislate for banana size, shape and even their curve so the workers must be sure there are no slip-ups with any fruit that aren't up to scratch. Any that fail are thrown onto the waste conveyor, but they aren't discarded. They'll be used to feed local life stock or sent to be processed into baby food.

Green bananas are then sent off for another bath which kills off any spiders who may be trying to hitch a ride to a supermarket near you. The fruit is given its trademark branding, and boxes are built to store them for their long journey across the Atlantic. This plantation has produced over 270,000 boxes of bananas. They are packed up and sent by the lorry load to the docks. Here they're put onboard ships in specially cooled containers for the journey to Europe. They won't be enjoying any sun on the deck though. It's an 11 day journey and being in the sun would ripen them too early.

When they are finally unloaded, the bananas are still green. This means they can be stored until the shops need them. They're stacked in sealed ripening rooms. When the stores need them, these rooms will be flooded with ethylene gas. This gas ripens the bananas at a regular, measurable pace which means the stores know exactly when they are ready for the shelves.

From the exotic fields of a Costa Rican banana plantation to the supermarket trolley in the UK, the Brits are certainly bananas about their bananas.

Did you know?

He was famous for his love of cheeseburgers, but another of Elvis Presley's favourite snacks was a fried banana and peanut butter sandwich.

枫糖浆

扫描二维码，观看中文视频。

一顿甜蜜的美式早餐不需要熏肉和鸡蛋，但需要薄饼。而如果没有枫糖浆，薄饼早餐是不完整的。那么，这种黏稠的蜂蜜替代品究竟来自何处？它是怎样被大批量生产出来，满足人们的好胃口呢？

枫糖浆毋庸置疑来自枫树。许多枫树生长在美国，但世界上最大的枫树林在加拿大。

巨大的枫树林被塑料管绑在一起。枫糖浆的旅程从这里开始。工人首先在树干上钻孔采集树液。用锤子把水管搋进树干，树液开始流入管道网络中汇聚。

春天是枫糖浆的收获季节。树木开始产生新鲜树液，为夏季长出新叶做准备。当树液通过根部上升为枫树提供营养时，工人安放的水管分流一些树液制作糖浆。每年可以从一棵大枫树上提取 190 升的树液，而不损害它的健康。

为了采集汁液，工人们想出了巧妙的方案。管道网络被精心布置成下坡方向。这样所有工作由重力完成。工人们密切关注管道网络，寻找泄漏或有问题的地方，如果一切顺利，汁液将最终来到收集棚。每天成千升的树液到达这里，但这是一个非常缓慢的过程。因为对糖浆的巨大需求量，种植园主安装了电动泵，径直把汁液从枫树中吸出来。在这个阶段，透明的树液很稀薄，看起来像水一样，虽然尝起来微甜，但没有传统的枫糖浆味道。

春季，这个安静的工棚变成热闹的蜂巢。采集人员按时光临，他们的工作是把树液从这里运到蒸发厂。要把枫树树液变成糖浆，需要加热到 100 摄氏度，使多余的水分蒸发。树液的主要成分是水，40 升树液才能提取 1 升糖浆，这就是它比其他类似品种更贵的原因。在巨大蒸发罐下烧起猛火，打开屋顶让水蒸气散发。一旦火焰升起，炉灶将被关闭。现在可以把树液倒进罐中了。

这是一个长期、缓慢的过程，但这不是员工可以放松的借口。凝结糖浆需要几小时，他们必须不断按时添火，使温度恰到好处。这看起来是非常简单的工作：烧火，看着水蒸发。实际上要更复杂，他们必须在紧要时刻密切关注树液。如有失误，一整批的糖浆就会完全报废。

蒸发使汁液变稠成为糖浆，但过多的熬煮会造成结晶。这将是灾难性的，整批的糖浆都必须扔掉。树液只有达到一定的浓度才能变成糖浆，工人用液体比重计进行测量。一旦达到合适的浓度，纯糖浆便通过箱底的龙头释放出来，现在缺少的只是一大沓热煎饼了。

随着屋顶打开，微风中荡漾着温暖的树液芳香，昆虫可能率先光顾和品尝。为了确保它们不会与新鲜糖浆一起装瓶，需要用棉布过滤。最纯粹、最优质的枫糖浆，从下面的龙头里流出来了。

虽然以相同的方式制成，枫糖浆根据颜色和质量不同分为几个等级。初春最早的枫糖浆

你知道吗？

在准备好莱坞大片《追梦女郎》中的角色时，超级巨星碧昂丝·诺尔斯用枫糖浆节食法减掉了 20 磅（1磅 =0.4536 千克）。

是透明的，有一种柔和微妙的味道。随着春意更浓，枫树产生的有机酸使糖浆颜色更重、味道更浓，倒在煎饼上仍然口感很好。最后一个等级是颜色最深、味道最重的，被用在工业烹饪中，或者制成人工香料和着色剂。最优质的糖浆被选送出来，装进传统的枫叶形状的瓶子里。你一旦尝过这种金黄色的糖浆，就会每天都找到吃煎饼的理由。

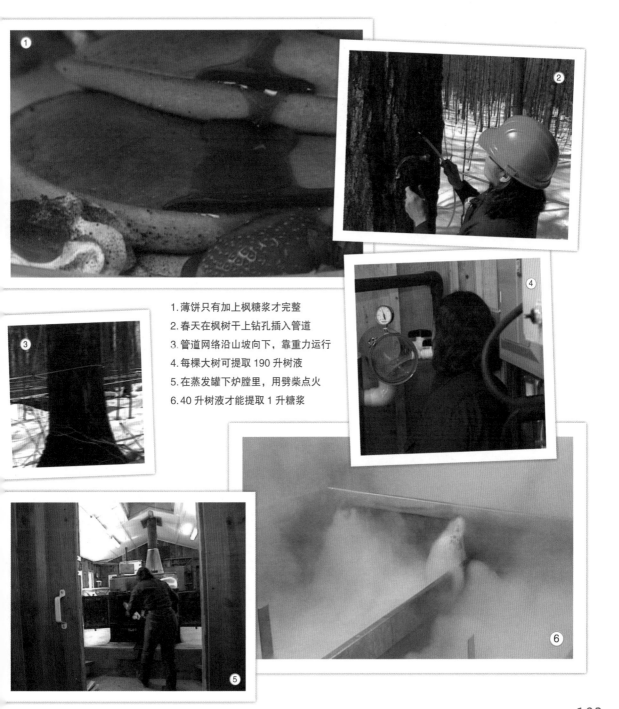

1. 薄饼只有加上枫糖浆才完整
2. 春天在枫树干上钻孔插入管道
3. 管道网络沿山坡向下，靠重力运行
4. 每棵大树可提取 190 升树液
5. 在蒸发罐下炉膛里，用劈柴点火
6. 40 升树液才能提取 1 升糖浆

7. 凝结糖浆需几小时不断添火
8. 水蒸气从屋顶散发
9. 用液体比重器测量防止结晶
10. 用棉布过滤防止飞虫混入
11. 枫糖浆按颜色划分等级
12. 优质枫糖浆装进枫叶形瓶子

Maple Syrup

扫描二维码，观看英文视频。

For a sweet American breakfast you don't need bacon and eggs, you need pancakes. But no pancake breakfast would ever be complete without maple syrup. So where does this sticky sweet alternative to honey actually come from and how is it produced in big enough quantities to satisfy a hearty appetite?

Maple syrup unsurprisingly is comes from maple trees. Many grow in the States, but the world's biggest maple forests are found in Canada.

Huge forests of these trees can be found tied together with plastic tubing. This is where the maple syrup begins its journey. It starts out as sap in the trees, which the workers collect by drilling holes into them. Taps are then hammered into place and the sap can start to flow into the network of pipes ready to be collected.

Maple syrups season is the spring. The trees begin to produce fresh sap to make new leaves for the summer. As it rises up through the roots to feed the tree, the workers taps divert some of it out to be made into syrup. Up to 190 litres can be taken out of a big tree each year without harming it's health.

To collect the sap, the workers have come up with an ingenious solution. The pipe network is carefully laid out to flow downhill. This way gravity does all the work. The workers do keep an eye on the network looking for leaks or problems, but if all goes well the sap should end up here in the collection shed. Thousands of litres of sap arrive here every day but it's a very slow process. Because there's a huge demand for syrup the plantation owner has installed a motorized pump to literally suck the sap out of the trees. At this stage the clear sap is thin and watery and although it would taste slightly sweet, it wouldn't have the traditional maple syrup flavour.

During the spring this quiet shed becomes a hive of activity. It gets regular visits from the collection guy who's job it is to take the sap from here to the evaporation plant. To turn maple sap into syrup, it needs to be heated to 100 degrees Celsius for the excess water to evaporate. Sap is mostly water and it takes 40 litres to get just 1 litre of syrup which is why it's so expensive compared to other varieties. Large fires are lit under the big evaporating tanks and the roof is opened so the water vapour can escape. Once the flames have properly taken hold the ovens are closed. The sap can now be released into the tanks ready to begin.

It's a long, slow process but that's no excuse for the workers to relax. To condense the sap into syrup takes several hours and they have to continually refresh the fire at regular intervals to keep the temperature just right. It looks like quite a simple job. Heat the fires. Watch the water evaporate. But it's more complicated than that. The sap has to be watched closely for the critical moment. If they miss it a whole batch can be completely ruined.

The evaporation thickens the sap into syrup, but too much cooking would crystallize it. This would be disastrous and the whole lot would have to be thrown out. The sap only becomes syrup when a certain density has been reached and the workers measure this using a hydrometer. As soon as it hits the right density the pure syrup is released via the tap in the bottom of the tank and all that's missing is a big stack of hot pancakes.

With the open roof and the smell of warm sap on the breeze, insects may have wandered in for a sample. To ensure they aren't bottled up with the fresh syrup, it's now filtered through cotton. What emerges from the tap below is pure finest quality maple syrup.

Although it's all made in the same way, there are several grades of maple syrup which are based on its colour and quality. The first sap of the season is clear and has a mild, delicate flavour. As spring progresses organic acids that the tree produces make batches darker and richer but it would still taste good on your pancakes. The final grade is the darkest and most intense and this is used in industrial cooking, and for artificial flavours and colouring. Some of the finest quality syrup is sent to be packaged up in the traditional maple leaf shaped bottles. And once you've tasted some of this golden syrup, everyday feels like a good excuse for Pancake day.

Did you know?

Whilst preparing for her role in the Hollywood blockbuster "Dreamgirls", R & B superstar Beyonce Knowles lost 20 lbs on a maple syrup diet.

甜菜制糖

白砂糖，英国每年吃掉超过 200 万吨。你可能不相信，在英国有一半的糖，是用这种不起眼的甜菜生产的，它和流行的紫甜菜同宗，糖用甜菜是种子长出来的。将种子涂一层营养物质促进生长，然后播种。

80 天后的结果是这样的。糖用甜菜在阳光下逐渐成熟，像任何其他的绿色植物一样甜菜靠光合作用生长。阳光被叶子吸收后转变成化学能产生葡萄糖和氧。

当甜菜长成后，巨大的拖拉机进行收获。从挖起到处理一条龙完成。第一步除去叶子。它们对食糖生产没有用处，因此被切掉。被翻出地面的甜菜汇集到收割机中。顺着引导管送进机器后，他们被推上后部的筐子里。这台收割机每小时能采集 25 吨以上甜菜，将它们装运到巨大的加工厂。英国农场每年生产近 700 万吨甜菜。林肯郡和埃塞克斯郡的土壤非常适合种植这种有用的作物。他们生产的糖将为茶、咖啡、蛋糕和点心带来甘甜。

然而在用于蛋糕或茶之前，首先需要好好洗个澡。清洗干净的甜菜从另一端出来，准备变成砂糖。接下来需要把大块的甜菜切成条，而不是送去进行油炸，这台机器把它们变小以便提取天然糖。成吨成吨的甜菜条送进这些大锅，热水开始工作了。烹煮甜菜条能将其中有用成分释放进水中。这些成分就是用来制糖的原料。

煮过的甜菜条也不会浪费，它们被压缩成小块，作为宠物食品。非常受兔子欢迎。工厂将用这些甜水制造砂糖，但首先要清洁处理。

水中含有多种成分会破坏糖的最终味道。加进石灰水，以滤除不需要的成分。纯净糖水通过过滤器，现在可以提炼了。

这样状态的糖水不是很黏稠，糖浓度相当低。为了让它变甜，这些混合物需要通过 6 个不同的锅炉。随着一个接一个通过锅炉，越来越多的水分被蒸发，最后留下黏稠的糖浆。糖浆是深褐色，看起来像冒泡的巧克力奶。

现在必须在很低的压力下再次煮沸，最后的沸腾过程让晶体从液体中析出。这就是我们都熟悉的糖。结晶的糖浆在离心机中旋转。褐色液体流出来，白色晶体留下来便可以采集了。

英国用这种工艺生产自身用糖的一半。另外一半靠进口，通常由甘蔗制成，它们大部分在加勒比和非洲种植。到达茶杯之前，结晶的白糖需要打包。装进标准的 1 千克袋子中。或者你更愿意用方糖。把糖湿润并压成小立方体。随着传送带向前移动，小推杆把方糖送出来。这些方糖排着整齐队列放进盒子中，准备供应英国全境的各类用户。感谢甜菜一族和大量的辛勤劳动，让你享受一杯又热又甜的茶。

你知道吗？

据估计，在每袋标准的 1 千克袋装糖中有超过 500 万个糖晶体。

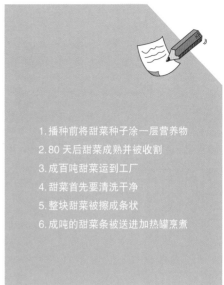

1. 播种前将甜菜种子涂一层营养物
2. 80 天后甜菜成熟并被收割
3. 成百吨甜菜运到工厂
4. 甜菜首先要清洗干净
5. 整块甜菜被擦成条状
6. 成吨的甜菜条被送进加热罐烹煮

7. 煮过的甜菜条压缩成宠物食品
8. 在烹煮完甜菜条的水中加进石灰水后滤除不需要的成分
9. 过滤后的糖水通过 6 个锅炉使水分蒸发
10. 低压下再次煮沸，沸腾过程让晶体从液体中析出
11. 糖结晶从离心机中送出
12. 准备包装的方糖

⑦

⑧

⑨

⑩

⑪

⑫

Sugar Beet Production

扫描二维码，观看英文视频。

White granulated sugar. Britain eats more than 2 million tons of the stuff every year. Now, you may not believe this but half of that is made in the UK using the humble sugar beet. Related to the popular purple beetroot, sugar beet starts out in a seed. They are coated with nutrients to improve growth and then planted.

80 days later and here's the result. The sugar beet plants are ripening steadily in the sun. Like any other green plant, the beet grows through the process of photosynthesis. Sunlight absorbed through the leaves is converted into chemical energy that produces glucose and oxygen.

When the beets are grown, massive tractors harvest the crop. They pick it and prepare it all in one go. First the leaves are removed. They are no use to sugar production so they're sliced off. Now the beets themselves can be dug out of the ground to be collected into the harvester. As they follow the guiding tube into the machinery they are forced up into the holding basket at the back. The harvesters can collect more than 25 tons of beet every hour. They're loaded up and sent on to the enormous processing plant. British farms produce almost 7 million tons of sugar beet every year. The soil in Lincolnshire and Essex is ideal for growing this useful vegetable. The sugar they produce will sweeten tea, coffee, cakes and desserts.

However, before they will go anywhere near a cake tin or a cuppa, they need a good wash. What emerges from the other side is freshly scrubbed beet ready to be turned into granulated sugar. Next, the big lumpy vegetables need to be cut down into chips. They're not going to be fried, but this machine makes the pieces smaller so the natural sugars can be reached. Tons and tons of beet chips are passed into these enormous boilers and the hot water gets to work. Cooking the chips helps release the elements from the beet into the water. These elements are what will be used to make the sugar with.

The left over chips aren't wasted. They are compressed into little nibble sized pieces and used as pet food. Apparently they are very popular with rabbits. To make sugar this sweet water is what the manufacturers want but first it needs to be cleaned. The water contains many elements which would ruin the final flavour. Lime water is added which filters out the unwanted material. The pure sugar water passes through the filter and can now be refined.

In this condition, it's not very thick and the sugar concentration is quite low. To sweeten it, the mixture is sent through 6 different boilers. As it passes through each one, more and more water is boiled off leaving a thick, sweet syrup by the end. The sugar syrup is a dark brown colour and actually looks like frothy chocolate milk.

It must now be boiled down again under very low pressure and this final boiling process creates crystals within the liquid. This is the sugar we are all familiar with. The crystallised syrup is now spun in a centrifuge. The brown liquid passes out leaving the white crystals behind which can now be collected.

Britain makes about half of its own sugar in this fashion. The other half is imported and is usually made from sugar cane instead. Most of this is grown predominantly in the Caribbean and Africa. To reach your tea cup, the crystallised sugar must now be packaged up. The standard 1 kilogram bag is filled and packed down. Alternatively, you may prefer to use sugar cubes. Raw sugar is moistened and pressed into cube forms. As the conveyor belt moves forward, tiny plungers force the new cubes out. These cubes can now be collected in neat rows and placed into a box, ready to fill the sugar bowls and tea rooms all over Britain. So thanks to the beetroot's cousin and a lot of hard work, you can now enjoy a hot, sweet cup of tea.

Did you know?

It's estimated that there are more than 5 million sugar crystals in a standard 1 kilogram bag of sugar.

收获鳄梨

扫描二维码，观看中文视频。

这是一只鳄梨，一种"多功能"水果，它在各类菜肴中使用，还有令人惊讶的额外好处——可以护肤。鳄梨价格昂贵，要说原因，因为获得它们是一项艰苦的劳动。世界上最大的鳄梨生产国是墨西哥，每年出产近百万吨。但我们的旅程从南非，另一个生产大国开始。

收获季节里，工人带着这些采集设备前往果园。每个杆子大约有 3 米长，在顶部有刀具和袋子。工人将袋子置于成熟的鳄梨下，从自己一端拉动绳子，操纵相连的刀具进行采割。切下的水果落入下面的袋子，工人在地面收集鳄梨。重要的是不要摔着它们，来自世界各地的买家只会收购完好的鳄梨。收获的鳄梨随后被送到巨大的加工厂。第一步是给它们好好洗个澡，以确保它们清洁和安全食用。

英国人对鳄梨的需求似乎无法满足。仅 1994 至 2004 年的 10 年间，进口量从每年 1.5 万吨增长到近 3 万吨，翻了一倍。

质检员密切注视生产过程，并除去任何有明显损伤或缺陷的鳄梨。西方消费者购物时，对水果应该是什么模样有非常具体的要求。所以必须保持高标准。达不到这个严格标准的鳄梨被送到回收箱但不会扔掉。它们被送到这里，巨大的料斗装着长得丑的或个头小的鳄梨。因为质量欠佳，它们被粉碎和煮开。然后果肉被挤压，机器生产出一种水和油的混合物。从水中分离出来的油是一种有价值的副产品，可以出售给化妆品行业人员。鳄梨油作为极佳的保湿剂有助于防止皮肤老化。

剩下的果肉也不浪费。一个守候在厂外的罐子将它们装上。每天由拖拉机运回种植园，成为鳄梨树理想的肥料。

只有最好的鳄梨被留在工厂里，工人现在可以进行分级和筛选了。分级的关键是鳄梨的重量，这家工厂根据它们尺寸大小分为 6 个不同等级。每个工人都要经手 14000 个鳄梨，这是个精细的工作。这一过程中对鳄梨造成伤害，仍然意味着必须将它抛弃。

在包装起来之前，还要做最后一个测试——由这台机器判断果实是否成熟。讽刺的是较软的水果反而被去除。在这个阶段鳄梨应该是硬的。它们还需要一段时间才能来到超市货架上。在大批量生产的现代世界，这类水果是在盒子里成熟，而不是在树上。一旦鳄梨准备好出厂了，工人就开始装箱。每个盒子要细心填满，这样鳄梨就不会在运输过程中移动和受到损害。装好的盒子每个约 4 十克，被送到一个巨型冰箱里。降温能延缓水果的成熟过程，就像你家里的冰箱一样。这样它们就不会在从非洲到我们超市货架的漫长旅途中熟过了头。数以千计的鳄梨货盘被装入巨大的冷藏室。与此同时，质量控制员检查水果被降温到大约 6

你知道吗？

世界上最大的鳄梨调味蘸酱是由加利福尼亚的一些啦啦队员制作的。他们为此使用了 36 升辣椒，2000 个西红柿和 6000 个新鲜的鳄梨。

摄氏度。他感到满意后，将叫来卡车装运水果货盘。

　　鳄梨被直接送上等候它们的船只，被运往欧洲、美国和世界其他地方，让人们享受鳄梨的美味。无论是切成片或加调料吃，这种异国水果理所当然地被端上餐桌，还得益于尽职尽责的工业生产流程。

1. 农民在采摘鳄梨
2. 鳄梨进入工厂第一步是清洗
3. 质检人员分拣出残次品鳄梨
4. 废弃鳄梨提取油脂后压成肥料
5. 对鳄梨进行分级和筛选
6. 每个工人每天处理 14000 个鳄梨

7.这台机器判断鳄梨是否成熟

8.工人准备为鳄梨装箱

9.鳄梨被送到巨大的冷藏室降温

10.鳄梨温度降至 6 摄氏度可装运出厂

11.大宗鳄梨准备出厂

12.鳄梨成为世界餐桌上的美味

Avocado Harvest

扫描二维码，观看英文视频。

This is an avocado, a very versatile fruit. It's used in a variety of cuisines and has a surprising bonus, it's good for your skin too. They can be expensive, and here's why ... harvesting them is hard work. The world's largest producer is Mexico where they grow nearly one million tons a year but our journey begins in South Africa, another big producer.

During the harvest season workers head for the fields armed with these collecting devices. Each pole is around 3 meters long and on the end is a bag with a cutter attached. The worker will place the bag under the ripe avocado and pull the string at his end, which activates the cutter. The freshly cut fruit falls into the waiting bag below and the worker can collect the avocados down at ground level. It's important not to drop them as the buyers from around the world will only buy undamaged fruit. The harvest is then sent to this enormous processing plant. The first step for the fruit is a good bath to make sure they're clean and safe to eat.

We just can't seem to get enough of the avocado here in the UK. In just 10 years from 1994 to 2004 imports doubled from 15,000 tons a year to nearly 30,000. The quality inspector keeps a close eye on the produce and rejects any that have obvious damage or problems. Western consumers have very specific ideas about how a fruit should appear when they buy it, so standards have to be kept high. Any avocados that don't come up to this strict standard are sent to the rejects bin, but they aren't thrown away. They're sent here. This enormous hopper contains all the ugly or under-sized fruit and because they're not good enough, they will be crushed and boiled. The pulp is then squashed and what emerges from the machine is a water oil mix. Once separated from the water, the oil is a valuable by-product sold to the cosmetics industry. Avocado oil acts as an excellent moisturizer and helps stop the skin from ageing.

The pulp that's left over isn't wasted either. It's passed out of the factory into a waiting tank. Everyday, tractors arrive and take it back to the plantation where it will act as the perfect fertilizer for the avocado trees.

With only the best avocados remaining in the factory, the workers can now get about the task of grading and separating them. The key to the grading system is the avocados' weight and this factory works with 6 different size groups. Each worker has to pack around 14,000 avocados, which is a delicate job. Damage at this stage still means the fruit must be thrown away.

Before it can be packed up, there's one last test to undergo. This machine determines whether the fruit is ripe yet. Ironically softer fruit is rejected. At this stage they should be hard. It will be sometime before they make it to supermarket shelves. In our modern world of mass production, fruits like these ripen in boxes and not on trees. Once the fruit is ready for shipping the workers start packing. Each box needs to be filled carefully so the fruit doesn't move about in transit and suffer some damage. The prepared boxes, which weigh about 4 kilograms each, are now sent to a giant fridge. Cooling the fruit slows down the ripening process like your fridge at home. This way it doesn't over-ripen during its long journey from Africa to our supermarket shelves. Enormous pallets of thousands of avocados are loaded into gigantic chilled storage containers. Meanwhile, a quality controller checks the fruit are chilled to about 6 degrees Celsius. When he's satisfied, the trucks are called in and the fruit pallets are loaded up.

They're then taken straight to the waiting ships to bring them over to Europe, the USA and everywhere else in the world where they enjoy avocados. So whether it's served sliced or sauced, this exotic fruit that's often taken for granted is delivered to the dinner table thanks to an exhaustive industrial process.

Did you know?

The world's largest Guacamole dip was made by some cheerleaders in California. They used 36 litres of chilli, 2,000 tomatoes and 6,000 fresh avocados.

有机草莓

美味的草莓。英国人喜爱这种小小的红色水果。如今草莓上市的季节远远比过去长，这要感谢大棚技术的使用，以及从日照时间更长的国家进口草莓。

有机作物的需求在激增，草莓也不例外。但有机草莓是怎么种出来的呢？

有机草莓的种植，始于有机肥料的使用，它们用牛和马的粪便制成。肥料分解释放出氮，对草莓的生长至关重要。

农民备好土地为下一季作物播种。遗憾的是不仅仅人类喜爱草莓，许多昆虫也喜爱它们，但对于有机农场主人，不能使用化学农药。现代工业杀虫剂对健康有危害，农民处理它们时，常常需要穿戴面具和防护服。这些化学物质会扩散到更广的环境中。有机农场用不同的办法，对付那些损害收成的害虫。农民用生物控制方法取代杀虫剂，引进自然的天敌，吃掉作祟的害虫。

农民面临的另一个问题是发霉，但能用天然的碳酸氢盐解决。类似用于烤面包使用小苏打，这种天然矿物质有助于草莓植株抵御感染。一旦作物开始生长，农民就往地里喷这种无毒的混合剂，帮助植物保持健康。

要让植物长出新鲜草莓，它们需要授粉。这一过程会自然发生，蜜蜂和昆虫能起促进作用，这也是为什么很多有机农场都有自己的蜂箱。如果农民使用化学农药，蜜蜂就难以存活。

有机植物给蜜蜂提供花粉，有助于产生果实。还有一个红利是蜜蜂采集的蜂蜜也能出售。当草莓成熟时它们被采集下来，准备送到包装厂。

有机作物往往比大规模生产的品种产量低，但有机草莓的需求没有减少。有机产品使农民在市场上卖到好价钱，对于他们采取环保种植方法，这是个良好的激励机制。许多客户愿意为健康、环保产品带来的放心支付更高的价钱。

作物在包装厂称重，扁盒子里的草莓将单独包装，为运输过程提供保护。它们将发送到欧洲各地的商店和市场，所以要精心包装。还要贴上清楚的标签，这样客户就能确切知道草莓来自何处。最后将它们重新装进盒子，像你每天在超市货架上看到的那样。无论你喜欢涂奶油还是蘸糖，有机草莓不仅环保也对你大有好处。

你知道吗？

传统的红色草莓只是这种受欢迎的水果中的一种，还有白色和黄色两种草莓，甚至有一种草莓像菠萝的味道。

1. 许多人喜爱草莓这种红色小水果
2. 大棚技术让草莓上市时间更长
3. 有机肥料用牛和马的粪便制成
4. 备好土地为下一季作物播种
5. 许多昆虫会危害草莓的生长
6. 引进自然的天敌吃掉害虫

7.喷洒无毒碳酸氢盐能抵御发霉

8.蜜蜂为草莓授粉

9.许多有机农场都有自己的蜂箱

10.为保证蜜蜂存活农民不使用农

11.有机草莓产量低但需求高

12.草莓包装称重运往市场

Organic Strawberries

The delicious strawberry. Britons love this little red fruit. Nowadays, the strawberry season is far longer than it used to be thanks to the use of poly tunnels and imports from countries where the sun shines a little longer.

Demand for organic produce has ballooned, and strawberries are no exception. But what really goes into growing an organic strawberry?

To start with the organic strawberry is grown using an organic fertiliser. This is made out of cow and horse manure. Its decomposition releases nitrogen vital for the strawberries growth.

After he's prepared his fields, the farmer will sow the seeds for next crop. Unfortunately it isn't only humans who enjoy strawberries. Many insects like them as well, but for the organic farmer, chemical pesticides are not an option. Modern industrial insecticides are so dangerous to the health that farmers often have to wear masks and other protective clothing when handling them. These chemicals can spread into the wider environment so organic farmers need a different solution to combat the bugs that threaten to harm the harvest. Instead of pesticides, the farmer uses biological controls, introducing natural predators that eat the insects infesting the crop.

Another problem the farmer faces is mildew, but he can tackle this using natural bicarbonate. Similar to the bicarbonate you might use for baking, this natural mineral helps the strawberry plants to fight off infections. Once the crops have started to grow, the farmer will spray the fields with this non-toxic mixture to help keep his plants healthy.

Now to get the plants to produce fresh strawberries, they need to be pollinated. This does happen naturally, bees and insects help the process along which is why many organic farms will have bee hives on them. Had the farmer used chemical pesticides the bees would not have survived. Organic production gives the bees a supply of pollen and helps produce the fruit. And as a bonus the bees also produce honey which can be sold. When the strawberries have reached maturity they're picked and collected ready to be sent to the packing plant.

Organic crops tend to have a lower yield than the mass-produced varieties, but demand for strawberries isn't going down. Organic produce will earn the farmer a premium on the market, which is a good incentive for them to turn to these environmentally friendly methods. Many customers will pay extra for the peace of mind a healthy, environmentally friendly product gives them.

At the packing plant the crop is weighed, and the punnets are then sent to be individually wrapped up to protect them on their journey. They will be sent to stores and markets all over Europe so they are carefully packaged. They are also given clear labels so the customer knows exactly where their strawberries have come from. Finally they're repacked into the cases that you see every day on the supermarket shelves. So whether you like them smothered in cream or just with a sprinkling of sugar, the organic strawberry is not only good for the environment, it's good for you too.

Did you know?

Traditional red strawberries are only one kind of this popular fruit, White and Yellow type exist, and one variety even tastes like pineapple.

橄榄油

扫描二维码，观看中文视频。

对于有健康意识的美食家来说，有种配料一定会让味蕾兴奋不已。从油炒到沙拉调料，橄榄油是每个现代厨房的必备品。虽然有多种级别和风味，但它们都有一个共同点。不，共同点不是装油的绿色瓶子而是橄榄本身。每种橄榄都具有独特的味道。通过不同种类橄榄的组合，制造商可以生产出独特口味的橄榄油。

生产过程开始于收获。传统的方法是用小耙子采集橄榄，但效率不高。像这个农场的大片橄榄树作物，就要代之以自动收割机了。机器围绕每一棵树，名副其实地"摇动橄榄枝"。新鲜的橄榄果被收集到下面的料斗里，连同一些树叶和小树枝——稍后很容易去除。这样一台机器在一个小时内采集的橄榄等于传统的农民一整天的采集量。当收获物送到生产工厂要先冲洗果实，以除去收集过程中混入的尘土、叶子或小树枝。残留下来更难处理的树枝被一个网架过滤掉，只让橄榄通过。

为得到最优质的橄榄油，橄榄应尽快压榨。传统的方法会让收获和研磨过程之间有一定的延迟。传统方法还使用这种轮状花岗岩大磨盘，将橄榄果和它的硬核一起磨成稠浆。但现代化生产仔细控制这两个重要因素。第一个是橄榄从收割机直接发送到磨床，其间的延滞被尽可能缩短。第二个是果实更温和的研磨，如果橄榄由重型磨盘磨碎，摩擦产生的热可能使橄榄失去风味。最受追捧的橄榄油被称为"冷榨"。如果橄榄果浆的温度超过 27 摄氏度就不能再称为冷榨，价值也随之降低。

是提取橄榄油的时候了。在传统的生产中橄榄浆层被夹在麻垫之间。一定量的橄榄浆被另一层垫子盖住，如此重复直到层层交替看起来像一摞巨型薄煎饼。这一摞垫子放到液压机中，把油从橄榄浆里榨出来。在下方收集到高质量的橄榄油带着传统的混浊的金黄色。现代的方法远没有那样剧烈，但同样有效。取代橄榄浆压榨的是在这样的离心机中旋转。在橄榄浆被转动时，橄榄油通过细筛孔流出，剩下残留的橄榄浆。然后将橄榄油吸出和存储。剩下的橄榄浆被收集到别处，但不会浪费，而会再循环用作肥料或动物饲料。

你知道后可能会惊讶，一旦榨出，橄榄油可以像葡萄酒一样品尝，体验明显地会非常不同，但这是专家用来评估产品纯度和风味的一种方式。纯的未经过滤的橄榄油被认为最有价值。如果残留物通过这样的离心分离机被过滤掉，生产出来的油就被认为略逊一筹。

你在超市货架上看到的标准等级，存放在这样的钢罐里。最后的残留物沉淀到底部，结果泵出来的就是大众市场上的透明橄榄油。

橄榄油通常装在绿色玻璃瓶里，这样做

你知道吗？

除了烹饪外，橄榄油还有多种用途——从保湿霜和按摩油到早期奥林匹克火炬的燃料。

有一个很好的原因。瓶子有助于过滤掉紫外线——它能招致橄榄油变质。最后,瓶子被密封并贴上标签,准备运到遍布世界各地的家庭和餐馆。

因此,无论是做调料的冷压初榨橄榄油,或标准等级的煎炒用油,橄榄油将在世界各地的厨房烹饪中长盛不衰。

1. 每种橄榄具有独特的味道
2. 传统采摘法用小耙把橄榄捋下
3. 现代农场大片橄榄树用收割机
4. 机器围绕每棵树把橄榄拍打下来
5. 清洗橄榄果并除去枝叶
6. 传统方法用石碾将橄榄果压磨

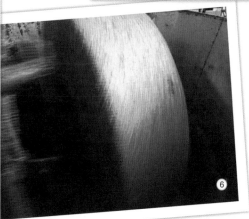

7.现代生产将橄榄直接从收割机送到磨床

8.冷榨橄榄浆温度在 27 摄氏度

9.传统生产中橄榄浆夹在层层麻垫间

10.一摞垫子在液压机中榨出油

11.现代生产从离心机中出油

12.橄榄油装入绿色瓶子以防紫外线

Olive Oils

扫描二维码，观看英文视频。

For the health-conscious gourmet, there's one ingredient that's sure to get the taste-buds sizzling. From frying to salad-dressing, olive oil is an essential in any modern kitchen. A wide variety of grades and flavours are available, but they all have one thing in common. No, it's not the green bottle it often comes in, but the olives themselves. Each type of olive has a unique taste. By combining different types the producer can create an oil with an individual flavour.

The process begins at harvest. The traditional, (time-consuming) method involves gathering and collecting olives using tiny rakes. But it's not very efficient. For a big crop like the one on this farm, an automated harvester is used instead. The machine surrounds each tree and literally shakes the olives off the branches. The fresh fruit is collected into waiting hopper along with some leaves and twigs, but these can easily be removed later. A machine like this can collect as many olives in an hour as it would take the traditional farmer to collect in an entire day. When the harvest reaches the production plant, the fruit is washed to remove any dirt, leaves or twigs that were caught up in the collection process. The more stubborn twigs and branches that remain are filtered out using a grill which only allows the fruit to pass through.

To get the best quality oil, the fruit should be pressed as soon as possible. Traditional methods mean there's a delay between harvesting and the grinding process. The original method also uses big millstones like these granite wheels to grind both fruit and the stones into a thick pulp. But modern production carefully controls two vital factors. First the fruit is sent directly from the harvester to the grinders with as little delay as possible. And second; the fruit is ground more gently, olives can lose flavour if the fruit is heated by the friction of heavy-duty grinding. The most sought after olive oil is called "cold pressed". If the fruit pulp goes over 27 degrees, it can no longer be called cold pressed and loses value.

It's time to extract the oil. In the traditional system, the pulp is layered between hemp mats. Each quantity of pulp is followed by another mat and so on, until the alternate layers look like a stack of giant pancakes. The stack is placed in a hydraulic press which literally squeezes the oils from the pulp. It's collected below and has the traditional cloudy, golden colour associated with good quality olive oil. The modern method is far less aggressive, but just as effective. Instead of crushing the pulp, it's spun in a centrifuge like this one. As it's spun round, the oil passes out through a fine mesh, leaving the pulp behind. The oil can then be siphoned off and stored, whilst the remaining pulp is collected elsewhere. It's not wasted though. It's recycled and can be used as fertiliser or animal feed.

It may surprise you to know that once it's been pressed olive oil can be tasted like wine. The experience is obviously very different, but this is one way the experts can assess the purity and flavour of a product. Pure unfiltered oil is considered the most valuable. If the residues are filtered out in a centrifuge like this one, what emerges is considered a slightly inferior grade of oil. The standard grade that you would find on supermarket shelves is stored in steel tanks like these. Here the final residues sink to the bottom and what's pumped out at the end is a clear olive oil for the mass-market.

Olive oil is usually bottled in green glass, and there's a good reason for this. It helps filter out harmful UV light that can cause it to deteriorate. Finally the bottles are sealed and labelled up, ready to be shipped out to homes and restaurants all over the world.

So whether it's cold-pressed extra virgin for dressing or standard grade for frying, olive oil continues to be cooked up in kitchens everywhere.

Did you know?

Besides cooking, olive oil has been used for a variety of purposes from moisturiser and massage oil to a fuel for early Olympic torches.

蘑菇

蘑菇，如果你以为它们都生长在森林的地面上，恐怕就需要再认识了。这里讲述一家工厂是如何大批量生产蘑菇的。

蘑菇的故事，从它赖以生长的堆肥开始。每天几卡车的堆肥运到这里，上面已经撒满了几百万个蘑菇孢子。再加入化肥，以利蘑菇生长。接着混合物被平铺到很大的托盘里。这个任务过去由人工完成，如今已被机械自动化取代，每15秒就可以装好一个托盘。这意味着和过去相比，种植人员可以生产出多得多的蘑菇，来满足巨大的市场需求。2005年，仅一个英国消费的蘑菇，就达到惊人的67000吨。最后，大量潮湿的覆盖物被撒到每个盘子表面。用以保护蘑菇孢子，并维持湿度以利它们更快生长。

为避免虫害，工厂采用加压环境杜绝害虫入侵。因为蘑菇生长不需要任何光线，工厂可以把托盘上下叠放来节约空间。它们被存放在这些巨大的环境控制房间里。温度和湿度都维持在蘑菇生长的最理想水平。

蘑菇受到定期检查，以确保按计划生长。在工厂里，蘑菇需要大约两周时间长成。这期间里它们天天都在变大，直到每个盘里的蘑菇多得快要溢出来。

装满成熟蘑菇的盘子被送到这里，心灵手巧的工人把奶油般颜色的作物采摘下来。他们每周收获可多达80吨。收获的蘑菇中，1/3会保鲜运到欧洲各地的蔬菜摊点和超市，并按照大小和重量分拣归类。剩下的蘑菇被用于熟食制作和比萨配料。

牛粪堆肥里长出来的蘑菇，要做的第一件事就是好好洗个澡。洗净之后，质检员挑出坏蘑菇，好的蘑菇进行第二次泡澡，它们根朝上或根朝下漂浮在水中。这意味着到达切片机的时候，都会被纵向剖开。此后切片蘑菇被冷冻，尽可能地保持新鲜。它们旅途的终点是这台机器。在这里，蘑菇被称重，然后倒进箱子里。封装的箱子经由传送带装上货盘。

这个工厂一年生产的蘑菇超过4000吨。看来我们对蘑菇的喜爱正越来越深。

你知道吗?

无论野生蘑菇或栽培蘑菇，采摘后都将继续生长。

1. 蘑菇是从工厂大量生产出来的
2. 撒满蘑菇孢子的堆肥运到工厂
3. 混入肥料后，堆肥被分装进大托盘中
4. 采用加压环境杜绝害虫入侵
5. 蘑菇无须光线，托盘可叠放
6. 托盘里的蘑菇迅速生长

7.托盘放在巨大的环境控制房间

8.工人将蘑菇采摘下来

9.对蘑菇进行分拣送往超市

10.挑剩的蘑菇送去清洗切片

11.堆肥中生长的蘑菇进行清洗挑捡

12.蘑菇切片后进行分装

Mushrooms

Mushrooms. If you think they all grow on forest floors you better think again. This is how a mushroom factory produces the fantastic fungi on a massive scale. This story of the mushroom begins with the compost it's grown in.

Lorry loads of manure are delivered here every day already laced with millions of mushroom spores. A fertiliser is added to help get them growing The mixture is then laid out into vast trays. This task used to be performed by hand, but automation has taken over and now a tray can be filled every 15 seconds. This means the growers can producer far more mushrooms than before to meet the enormous demand. In 2005 we ate a staggering sixty seven thousand tons of mushrooms in the UK alone. Finally a layer of damp mulch is spread liberally over each tray. It protects the mushroom spores and keeps the humidity in so they can grow faster.

To keep the crop free from insects they use a pressurised environment that keeps the bugs at bay. Because mushrooms don't need any light to grow the factory can save space by stacking the trays on top of one another. They are stored in these gigantic climate controlled chambers where the temperature and humidity can be kept at the ideal level for mushroom growth.

The crop is regularly inspected to make sure things are going to plan. In this factory the mushrooms take approximately two weeks to reach maturity and during this period they grow in size everyday until each tray is almost over-flowing.

The trays with the ripened mushrooms are taken here, where nimble-fingered workers pick the cream-white crop. They harvest up to 80 tons per week. A third of the harvested mushrooms are shipped fresh to vegetables stands and supermarkets all over Europe. Here they are sorted according to their size and weight. The remaining mushrooms are destined for ready meals and pizza toppings.

Having grown in cow-manure compost, the first thing the mushrooms need is a good bath. Once they are clean a quality controller picks out any damaged ones and then the healthy mushrooms go into another bath which makes them float either stem up or stem down. This means when they hit the slicer they're all cut vertically. They're then sent off to be frozen ensuring they'll stay as fresh as possible. The final leg of their journey takes them to this machine, where they are weighed and then dumped into boxes. The sealed boxes are taken by conveyor belt to be loaded up onto pallets.

With this factory producing over 4,000 tons of mushrooms each year. It looks like mushrooms are really growing on us.

Did you know?

Both wild and cultivated mushrooms continue to grow after they are picked.

樱桃果酱

在整个英国和其他一些国家，你都会在早餐桌上看到面包和果酱。但人们是如何把美味的黑樱桃变成满瓶的果酱呢？

从逻辑上讲，应该始于种植樱桃的果园。当樱桃完全成熟，农夫拿出了这个巨大的垫单。看来他要把垫单铺到树下，再把所有的樱桃摇下来。事实上，他正是这样做的。

这个机械手臂连接到树干上开始工作。所有成熟的果子都落到下面的垫单上，很容易被收集起来。树上留下的只是空果茎。农民要做的就是开启发动机，把垫单子卷回去，千百个成熟的樱桃落进采集盘。多简单！

农民装满一箱又一箱的新鲜樱桃，准备运到加工厂。

首先把混在其中的叶和茎除去。这个敏捷的滚轮系统能抓住细茎，把它们拉出来。只有樱桃能通过滚轮，到达制作过程的下一步。

接下来樱桃被冲洗，但有个东西水洗不掉——中间的核。下面这个奇妙装置抓住每一颗樱桃，送进脱核机。尖刺穿过樱桃取出坚硬的不能吃的核，留下纯净新鲜的果肉。樱桃终于准备变成果酱了。

果酱含糖，这不会令人惊讶。但知道含糖量多少可能让人惊讶。比如制作这种樱桃果酱，每千克水果需要用 1 千克糖。把糖加进水果中，在这些大铜锅里烹煮。下一个配料是果胶，它让液体黏稠变成果冻状。混合物就有了人们熟悉的果酱形态。将它和柠檬汁同时添加到水果里，混合物进一步烹煮。15 分钟后厨师将取样，在装瓶前检验特定的糖含量和均匀度。热的水果糊倒入漏斗，被带到楼下的装瓶厂。

在工厂的这一部分，刚清洗过的玻璃罐从龙头下通过，定量的果酱被装进去。果酱装罐后直接送到封盖机。每个瓶子加盖时，热蒸汽也被封了进去，这样做有个好理由。罐子关闭后直接送到冰箱。当快速冷却时，罐内产生真空将盖子封紧。如果不开启的话，每瓶果酱能保质 2 年以上。

最后，把洒溅出的果酱从罐子上洗掉，然后送去贴上传统的商标。从果树到工厂，再到手边的一片面包……像樱桃酱这样做成罐头的水果，确保离开果树后仍然能够被长久享用。

你知道吗？

有些果酱的食谱可以追溯到罗马时代。 但他们没有使用糖来使果酱变甜，而是使用了蜂蜜。

1. 樱桃果酱是早餐必备的美味
2. 成熟的黑樱桃等待采摘
3. 铺好垫子，将机械臂连接树干
4. 机器臂摇动树干，樱桃落下来
5. 垫子卷回将摇落的樱桃采集
6. 樱桃洗净并去除杂质

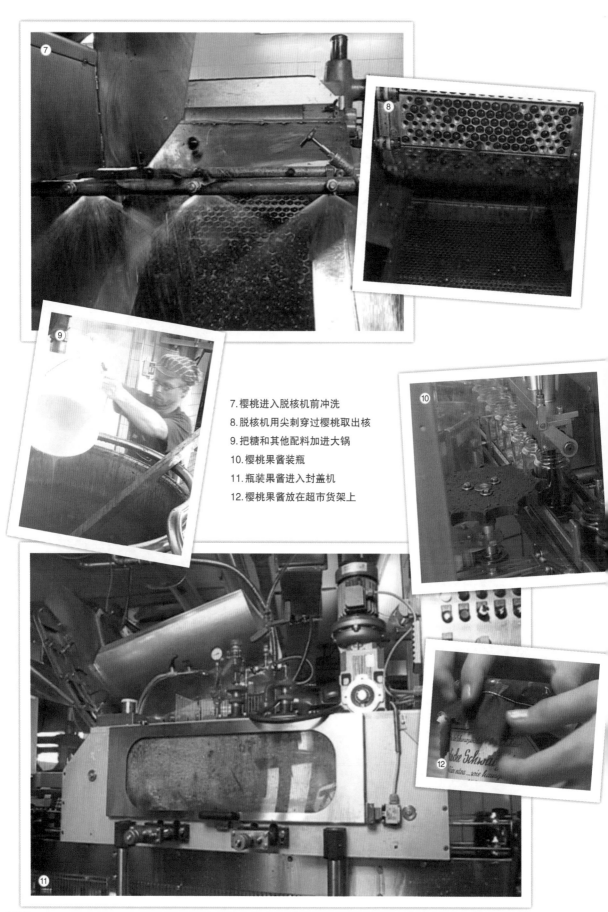

7. 樱桃进入脱核机前冲洗
8. 脱核机用尖刺穿过樱桃取出核
9. 把糖和其他配料加进大锅
10. 樱桃果酱装瓶
11. 瓶装果酱进入封盖机
12. 樱桃果酱放在超市货架上

Cherry Jam

Throughout the UK and plenty of other countries besides, bread and jam is one thing you'll probably find on the breakfast table. But how do you turn delicious black cherries into a jam jar full of well, jam?

The logical place to start would be the orchard where the cherries are growing. When they reach full ripeness, the farmer brings out this gigantic sheet. Now it looks like he's going to spread it under the tree and just shake all the cherries out. In fact, that's exactly what he's going to do.

This mechanical arm is attached to the tree and does its work. All the fruit that is ripe, will fall into the waiting sheet below where it can easily be collected up. All that's left behind are the empty stems. Then all the farmer has to do is turn on the engine, and the sheet rolls itself back up, depositing thousands of ripe cherries into the collection tray. Simple.

Farm workers collect box after box of fresh cherries ready to take them to the processing plant.

The first step is to remove all the leaves and stems that are still mixed in with the fruit. This handy roller system traps the stems and yanks them off. Only cherries make it past the rollers to the next step of the preserve process.

Next the cherries are rinsed, but there's one part the water can't remove. The stone inside. This next contraption catches each individual cherry. They are then passed up into the stoning machine where spikes pass through the cherry removing the hard inedible middles. All that's left behind is pure fresh fruit. The cherries are finally ready to be turned into jam.

Now it won't come as a surprise to learn that jams contain sugar, but it may surprise to learn just how much. To make this cherry jam for instance, for every kilo of fruit you use, you need 1 kilogram of sugar too. The sugar is added in with the fruit which is cooking away in these big copper bowls. The next ingredient is pectin. It will help thicken the liquid into a jelly-like substance giving the mixture that familiar jam-like texture. This will be added to the fruit along with lemon juice and the whole combination is cooked down further still. After 15 minutes, the chef will take a sample. The sample is checked for a specific sugar level and consistency before it is ready for bottling. The hot fruity preserve is poured into the waiting funnel which will take it to the bottling plant on the floor below.

In this part of the factory, freshly cleaned jars pass under the tap continually and a measured quantity of the jam is released. As soon as it's in the jars, they're sent directly to the capping machine. Hot steam is included as the top is put on each jar and there's a good reason for this. After they've been closed, the jars are sent straight to the refrigerator. As they cool quickly, a vacuum is created inside the jar which seals it tight. If it's unopened, each jar of jam will stay fresh for up to 2 years.

Finally any stray jam is washed off the jars and they are sent on to receive their traditional labels. So from the the tree to the factory to a slice of bread near you, preserving fruit in this way ensures that treats like cherry jam can be enjoyed long after the fruit has left the trees.

Did you know?

There are recipes for jam that date back to Roman times. But instead of using sugar to sweeten them, they used honey.

巧克力

有些人喜欢黑色，另一些人必须要牛奶味，甚至还有一些想要白色。有一件事许多人都相同，这就是对巧克力的喜爱。这种深受欢迎的零食有如此多的品类，以至不可能一一列举。但无论形状、大小或夹着什么，巧克力是一种世界性爱好。

不管在普通人嘴里咀嚼或在专业的测试人员口中溶化，味道都非常重要。从标准的牛奶巧克力条，到充满榛果的异国巧克力块，最重要的成分是巧克力本身。也许有人会感到惊讶，它竟然是长在树上的。

可可树生长在赤道附近热带的高温气候地区，至关重要的可可豆长在树上的豆荚中，这就是巧克力生产商所要寻觅的。每个豆荚含有40～50粒可可豆，饱含脂肪和蛋白质。这个数量能生产足够的可可用来做一块巧克力条。可可豆放在阳光下晒干，然后被运到世界各地的大型巧克力生产公司。

可可豆到达时，要做的第一件事是检验。约100粒样品可可豆放到特制的切刀里被切成两半。任何发霉或受昆虫损坏的批次将被拒收。一旦测试通过，成千上万的可可豆被直接送到烤炉中。

这就是提取可可的地方。可可豆烘烤后被磨碎。这将产生油状的液体，富含脂肪、可可和蛋白质。但在这个生产阶段你不会想吃它。它尝起来很苦。要把这些黏糊转化为甜巧克，需要在混合物中加进奶粉和糖并充分搅拌。产生的黏稠物看上去更像巧克力，尝起来味道也更好，但仍不具有巧克力条的稠密度。为了得到光滑的质感，首先巧克力要再次研磨。这能

在混合物中形成晶体，这样做的重要性后面会清楚表明。研磨过程经过几次重复，直至稠密度适当。质地达标后整批巧克力被送进这些庞大的加热搅拌机。在这里被加热到80摄氏度左右，并持续24小时。

现在加入一些坚果。这种巧克力条中用的榛果从土耳其一路运来。在它们加进巧克力之前，工人必须筛选出次品。好的榛果被送到装满巧克力的缸里裹上一层，所有混合物送到最后生产线。这个机器排出的设定分量正好做一条巧克力。

每个模子里装进的熔化物含有大约3/4的巧克力和1/4的坚果。机器的不断振动可以除去气泡并让巧克力和坚果混合物沉降到模子中。

除了味道，吃的体验中另一个重要因素，是巧克力条发出令人垂涎的断裂声，秘诀在于我们前面看到的结晶过程。当巧克力被加热时，烹煮过程去除了大部分晶体，但一些耐热的晶体例外，这里显示为黄色。正是这些晶体，使巧克力在折断时发出那种特色的破裂声。

除了水果和坚果，还有一系列各种风味的填料。生产它们需要一个非常不同的过程。首先将液体巧克力倒入模子。接着也许令人惊讶，模子

你知道吗？

最大巧克力块的世界纪录为 2.28 吨，但新加坡的一家生产商已经制造了重达 2.688 吨的"挑战者"。

又被直接翻过来。通过这样做，厨师让每个模子内涂上一个巧克力层，并为填料留下空间。随后是装进。填料可以有各种风味，从奶油或者酸奶，到太妃糖或者焦糖。把填料放进模子后，在顶部喷上最后一层巧克力。它将成为新做出来的巧克力条底部。多余的部分被刮掉，留下来巧克力条进行固定。一旦巧克力硬化了，就被锤子从模具中敲出来，它们可以出厂了。

如果你是个巧克力迷，将会很高兴听到，仅仅这家工厂每天就可以生产250万条巧克力。那是你喜爱的250吨巨量甜点。有这么多巧克力要包装，很容易理解为什么它们现在要加速奔向生产过程的下一站。这个包装机每分钟能包装329条巧克力，人眼很难跟上。一旦包装完毕，它们将被发往全球超过65个国家。

关于吃巧克力的利弊得失，互相冲突的报告一直存在，但如果古老的格言可以相信，吃点你喜爱的东西对你有好处。

1. 可可豆长在树上的豆荚里
2. 每个豆荚含有 40 ～ 50 粒可可豆
3. 可可豆晒干后抽样切开质检
4. 可可豆磨碎产生油状的液体
5. 加进奶粉和糖搅拌至黏稠适度
6. 对榛果进行筛选

7.巧克力加进榛果

8.榛果巧克力倒入模具

9.夹心巧克力先做出外壳

10.夹心巧克力加进填料

11.巧克力准备包装

12.巧克力被普遍喜爱

Chocolate

扫描二维码，观看英文视频。

Some desire dark, for others milk is a must, and there are even those who want it white. There's one thing many people have in common and that's a love of chocolate. There are so many variations of this popular treat that it would be impossible to list them all. But whatever the shape, size or filling, chocolate is a universal favourite.

Whether it's munched in ordinary mouths or dissolved on a discerning taste tester's palate, the flavour is what's important. From a standard bar of dairy milk to an exotic ingot filled with hazelnuts, the most important ingredient is the chocolate itself and it may come as some surprise that it actually grows on trees.

Cocoa trees grow in hot tropical climates around the equator and it's the all-important beans that grow inside large pods on the trees that the chocolate chefs seek. Each pod contains between 40 and 50 cocoa beans full of fat and protein. This amount can produce enough cocoa to make a single bar of chocolate. The beans are left in the sun to dry and then shipped to the large chocolate production companies around the world.

When they arrive, the first thing that's done is a test. A sample of about 100 beans is placed into this custom guillotine, where they are sliced in half. Any mould or insect damage and the batch will be rejected. If they pass the test thousands upon thousands of beans are sent directly to the roasting ovens. This is where the cocoa is extracted.

Once the beans have roasted, they are ground. This releases a rich oily liquid made up of fat, cocoa and proteins, but at this stage of the process you wouldn't want to eat it. It's very bitter. To turn this thick sludge into sweet chocolate, milk powder and sugar are added into the mixture and stirred thoroughly. The resulting thick mass looks more chocolatey, and would taste better too, but it still doesn't have that chocolate-bar consistency just yet. To get that smooth texture, first the chocolate is ground up once more. This forms crystals in the mix, and the reason these are important will become clear later. The grinding process is repeated several times until the consistency is just right. When that critical texture has been achieved, the whole batch is sent to these enormous heated mixers. Here it's cooked to around 80 degrees Celsius for 24 hours.

And now for some nuts. The hazelnuts used in this bar have come all the way from Turkey. But before they are added to the chocolate, workers have to filter out the rejects. The good nuts are added to a vat full of chocolate and given a thorough coating before the whole combination is sent to the final production line.

This machine pours out a set measure for each bar. Each form holds a melted mass which is about 3/4 chocolate, 1/4 nuts. The machine vibrates continually to remove unwanted air bubbles and help the chocolate-nut mixture to settle into the form.

Apart from taste another key part of the eating experience is the mouth-watering cracking sound of a breaking bar. The secret lies in the crystallising process we saw earlier. When chocolate is heated, the cooking process removes most of the crystals, except for the heat-resistant variety, seen here in yellow. It's these crystals that create that characteristic crack when a block is broken off.

As well as fruit and nuts, there are a range of varieties with flavoured fillings. A very different process is needed to make these. First the liquid chocolate is poured into forms. But then perhaps surprisingly, it's tipped straight back out again. By doing this, the chefs give each form a chocolate coating, leaving room for the filling. This comes next. It can be anything from flavoured creams or yoghurt, to toffee or caramel. Once that's settled into the forms, a final layer of chocolate is squirted on to the top, or what has now become the bottom of the new chocolate bar. Any excess is scraped away and the bars are left to set. Once they've hardened up, they're knocked free from the moulds by hammers, and they're ready to go.

Now if you're a chocaholic, then you'll be very pleased to hear that this one factory alone produces 2.5 million bars every single day. That's a massive 250 tons of your favourite sweet snack. With that much to pack up, it's easy to see why the chocolate is now accelerated to go through the next stage of the process. This packing machine can wrap 329 bars a minute which is pretty hard for the human eye to keep up with. But once they're all wrapped up, they'll be sent to over 65 countries worldwide.

There have been conflicting reports about the pros and cons of eating chocolate, but if the old adage is to be believed, a little of what you fancy does you good.

Did you know?

The world record for the largest chocolate bar stands at 2.28 tons, but a producer in Singapore has made a challenger weighing 2.688 tons.

速溶咖啡

扫描二维码，观看中文视频。

每天，全世界的人要喝掉 22.5 亿杯咖啡。人们喝新鲜咖啡的历史长达 500 年，但速溶咖啡则是"街头顽童"，直到 1901 年才被发明。

每天 70 吨的新鲜咖啡豆从南美运到这个速溶咖啡工厂。这个巨大的筒仓能储存多达 560 吨咖啡豆。生咖啡豆被过滤到大烤箱里，在 200 摄氏度的温度下烘烤。咖啡豆在这里呈现了独特的颜色。它们被不停地搅拌以确保受热均匀，不会烤焦。

烤好的咖啡豆落入一个工业碾碎机，在那里被磨成粗粒粉末。粗粉粒会掉进 8 个巨大的咖啡机之一。其中的高温高压蒸汽将"味道"提取出来。它们做出的新鲜咖啡足够让一支军队数量的上班族清醒起来。仅 1 小时就生产出 18000 升。

咖啡被加热，直到浓缩成黏稠的精华后，铺展到传输带上前往下一站—— 冷冻厅。工人们必须裹上保暖的衣服来抵御这里如同北极的气温——零下 50 摄氏度。要锁住咖啡的香味，浓缩物必须冷冻坚实。最少要达到零下 40 摄氏度。在这条 30 米长的传送带的尽头，冻结好的浓缩物被打碎成颗粒。如此严酷的环境里，工厂使用摄像头来监控生产过程是否顺利。如果工人在冷冻厅里待得太久会被冻伤。

深冻的颗粒中仍然含有水分需要去除。它们被放入托盘，用 5 小时通过一个低压管道。

这里所面临的挑战，是既要将水分去除又不能让咖啡变回液态，否则香味就会逃逸散发。以下是工作原理：咖啡颗粒在真空中被 60 摄氏度的温度加热。冰冻的水分在负压下直接变成水蒸气。这个过程叫作升华。

颗粒从真空环境中出来的时候，已被冷冻干燥完毕。颗粒锁住了咖啡的香味，并且在室温下保持固态。将托盘翻转过来，颗粒落入巨大的储存袋，准备包装。

瓶子沿着传送带移动，在不到 1 秒的时间里就被装满。经过密封后，朝着工厂大门前进，在途中被贴上标签。每周有 420 吨速溶咖啡离开这家工厂，成为我们每天要喝掉的 22.5 亿杯咖啡中的一小部分。✎

你知道吗?

咖啡是世界上交易最多的食品或饮料商品。

1. 每天 70 吨鲜咖啡豆从南美运来
2. 生咖啡豆被送到大烤箱里烘烤
3. 烤箱里的咖啡豆不停翻动

4. 碾碎机把咖啡豆磨成粗粒粉末
5. 复杂的咖啡加工设备
6. 咖啡加热浓缩后铺展到传输带上

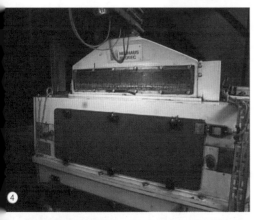

7. 工人进入零下 50 摄氏度的冷冻室

8. 冻结咖啡被打成颗粒

9. 深冻咖啡颗粒仍需脱水处理

10. 低压管道中咖啡颗粒去除水分

11. 咖啡颗粒落入袋中准备包装

12. 成品咖啡源源不断出厂

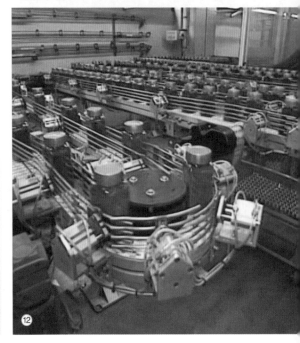

Instant Coffee

扫描二维码，观看英文视频。

Across the world we drink about two and a quarter billion cups of coffee every day. We've been drinking it fresh for over half a millennium but instant coffee is the new kid on the block. As it wasn't invented till 1901.

70 tons of fresh coffee-beans arrive from South America at this instant coffee factory every day. Up to 560 tons are stored in this huge silo. The raw beans are filtered down into large ovens where they're roasted at 200 degrees, this is where they get their characteristic colouring. They're stirred constantly to makes sure that they're evenly roasted without burning.

From there the beans fall down into an industrial mill where they're ground into a coarse powder.This plummets down to one of eight huge coffee machines where the flavour is forced out by hot steam and pressure. They brew up enough fresh coffee to wake up an army of office workers. Eighteen thousand liters in just an hour.

The coffee is heated until its condensed into an extract and that's spread on to a convey belt which will take it away to the next station, the freezing hall. The workers have to wrap up in thermal clothes to protect themselves from the arctic temperatures- minus fifty degrees Celsius. To lock in the coffee's aromas the extract must be frozen solid. At least as cold as minus forty degrees. At the end of a 30-meter-long conveyor its broken up into granules. In the harsh environment, they use cameras to check everything is running smoothly. If a worker spent too long in the deep freeze he would be going home with frost bite.

The deep-frozen granules still contain water which they need to get rid of. Stacked up on trays they're driven through a low pressure tube for five hours. The challenge is to remove the remaining water without the coffee becoming liquid again as this will release the aromas. Heres how it works. The granules are heated at 60 degrees Celsius in a strong vacuum. Under pressure the frozen water vapourises and turns straight into steam. This is a process called sublimation.

As the granules come out of the vacuum they have been freezed dried. The aromas have been locked in and they will stay solid at room temperature. The trays are flipped over and granules fall into huge storage sacks ready to be packaged.

The jars move along a convey belt and are filled in less than a second. They get an air tight seal and move towards the factory door, picking up a label on the way. 420 tons of instant coffee leave this factory every week. But it will be made up into a small proportion of those two and a quarter billion cups of coffee we drink each day.

Did you know?

Coffee is traded food or drink commodity in the world.

袋装茶

有的时候什么都比不上一杯好热茶。现在我们大多数人饮用从超市买来的袋装茶，但茶叶的生产却始于千里之外。印度是世界上最大的茶叶生产国之一，茶叶行业在这里雇用了将近150万人，在茶叶种植园里，工人从三月到十月之间采摘宝贵的茶叶。沏出一杯茶，最好的部分是叶子尖端和上部的嫩叶。

茶叶需要在新鲜时进行处理，以保证最佳的质量，这一切都要就地完成。首先，茶叶被碾压，将茶叶打烂，并释放出茶叶味道的汁液。下一个步骤是发酵。破碎的叶子在潮湿环境中铺在桌子上，使得茶叶能吸收氧气，变成铜棕色。接着，茶叶通过一系列的筛子。不同的茶叶尺寸决定了泡茶需要多长时间。

茶叶做好了，可以装盒并送到世界各地的工厂，包括汉堡。来自印尼、中国和印度的茶叶来到这里，准备由茶叶专家进行混合，做出受欢迎的品种。

在试验室，品茶师每天从各批次货物中取样，准备好数百杯茶。她仔细称出完全相同分量的每种茶叶以便准确的进行比较。她还需要使用同量的水，用相同的时间泡茶，正如茶叶爱好者所知，所有这些因素都会影响一杯好茶的味道和浓度。大约6分钟后滤掉茶叶，将它们放置一旁做后续检查。这位女士一定能分辨出阿萨姆茶和大吉岭茶，要成为品茶师，你必须至少接受5年训练！

首先检查叶子，她通过香味、大小、颜色和脆度来做出判断。这些信息告诉她茶叶的质量怎样，它们的干燥程度如何，最终能否泡出一杯好茶。然后该品尝滋味了。她快速把茶啜进口中，同时深吸一口。这会让茶充进气体变成细雾，接触到她所有的味蕾。

当她做完了评估后，工厂从不同种类的茶叶中制作出招牌产品，这时将装成茶包，最后贴上标签。所以，下次你在享受一杯茶的时候，想一想那些品茶师们吧，他们品茶之后得把茶吐掉，因此你才不需要这样做。

你知道吗？

茶占英国人每日液体摄入量的40%。

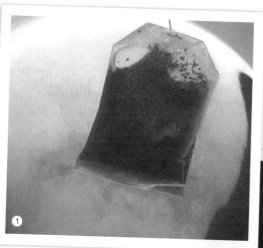

1.冲泡袋装茶

2.3 月到 10 月是印度采茶季节

3.叶芽才能制成最好的茶叶

4.把叶子碾压磨碎

5.破碎的叶子铺在桌子上发酵

6.茶叶通过一系列筛子

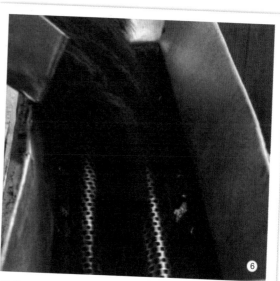

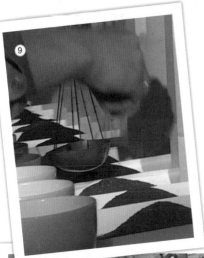

7.德国汉堡有大型袋装茶工厂

8.印尼、中国、印度的茶叶来到这里

9.精准称好茶叶分量

10.品茶师鉴别各种茶叶的品质

11.茶叶包装生产线

12.袋装茶准备出厂

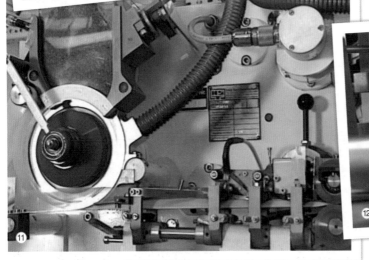

Tea Bags

扫描二维码，观看英文视频。

Sometimes you just can't beat a nice hot cup of tea. Now-a-days most of us make a brew with tea bags brought in the local supermarket, but tea starts life many thousands of miles away. India is one of the world's largest tea producers. The industry here employs nearly one and a half million people and in tea plantations like this one workers pick the most valuable leaves between March and October. The best parts of the plant for making a cuppa are the tip and the top leaves.

These need to be processed whilst still fresh to ensure the best possible quality so it's all done on site. First the tea leaves are rolled. This breaks them down and releases the juices that give tea it's flavour. The next step is fermentation. The broken leaves are laid out on to tables in a humid environment. This allows them to absorb oxygen and they turn a copper brown colour. The tea is then sorted through a series of sieves. The different sizes affect how long the tea will need to be brewed for.

Now the tea is preserved it can be boxed up and sent to factories around the world; including here in Hamburg. Tea from Indonesia, China and India arrive here ready to be blended into popular varieties by tea experts.

In the testing room, a taster prepares hundreds of cups of tea everyday made from samples of each shipment. She carefully measures out exactly the same quantities of each varieties so she'll be able to compare them accurately. She also needs to use the same quantity of water and brew them for the same amount of time, as tea loves know all of these factors affect the flavour and strength of a good cuppa. After about six minutes she strains off the tea leaves and puts them to one side for further examination. This woman definitely knows her Darjeeling from her Assam, to become a taster you have to train for a minimum of five years!

First she inspects the leaves, she's judging them by their fragrance, size, colour and crispness. These tell her how good quality the leaves are. How well they've been dried and ultimately if they will make a great cup of tea. Then it's time for her to get slurping. By quickly slurping the tea into her mouth she also takes in a healthy gulp of oxygen. This aerates the tea into a fine mist which travels over all of her taste buds.

When she has made her assessment the factory then makes it's house blend from variable quantities of the different teas. This is then packed up into tea bags and finally labelled. So the next time you're enjoying a fine brew think of all the tasters out there spitting out tea so you don't have to.

Did you know?

Tea accounts for 40% of the UK's daily fluid intake.

芦笋

它在欧洲各地被当作一种美味。通常只配着黄油吃。白芦笋是一种昂贵的蔬菜，生产它需要大量劳动。一切从它生长的土壤开始。这里需要大量的准备工作。

芦笋喜欢含有丰富肥料的疏松泥土。农民要非常精心为这些纤弱植物准备生长的地方。用专门的拖拉机翻动土壤，以确保足够疏松。只有当土壤疏松程度达到要求后，农民才能播下种子，但还不是种芦笋。

芦笋需要一年时间长出幼苗，但并不能收获。幼苗是芦笋根束或根冠成长过程的第一阶段，为最好的芦笋收成打下基础。

当时令到来，根冠被挖出来，准备再次栽种。它们被埋进很深的地下，远超过自然深度。芦笋茎要长到地表寻找阳光，深植使芦笋茎比一般情况下长得更长。这种幼茎便是美味的白芦笋。它是白色的，因为没有见过阳光。绿芦笋已经晒过太阳，因此含有更健康的维生素和矿物质。但这里种植的白色品种被认为更美味，农民种植它能得到更高的收入。

要为芽苗生长获得额外的深度，农民在每一行作物上堆积更多的土。随后将每一行用黑色篷布覆盖，以增加土壤温度并促进成熟过程。芦笋的季节是5月至6月，所以时间很短。通过提高温度，农民延长了生产时间和收获季节。换来更大的利润。

时机成熟时，芦笋用手工采摘。对幼茎逐个检查，已经成熟的被挖出来。但每一个根束可以长出几只茎，所以采摘者再次把它小心掩盖，以便多生长些时间。

娇弱的芦笋茎采集到篮子里，并送到农舍加工厂。芦笋是一种昂贵的食品，因为需要两年生长，一辆吉普车就足以把当天的所有收获带回农家了。

下一步是为用户备好芦笋。通常从清洗开始。一旦洗干净就经过铡刀，在这里切去底部。芦笋是一种精致的食品，餐馆和商店老板希望他们的客户有好的体验，除了味道之外也情愿为芦笋的外观付钱。新切好的芦笋再清洗一次，然后送去分级。农民得到的价格取决于芦笋的粗细、外观和颜色。芦笋从一道坎上滚过，相机纪录下它们的运动。按照标准进行分级。

分拣系统会把它们投进相应的盒子里，工人再次检查每个芦笋。任何有损伤或污垢的都会除去，然后芦笋准备送去冷藏。

将每筐芦笋浸入冷水用冰覆盖，能中断芦笋的成熟过程，保持新鲜直到出售。这些筐子放入冷库，直到农民为顾客发货时才离开那里。

能够用来作为配菜，或黄油烹饪。这种备受欢迎的美味叫白芦笋。

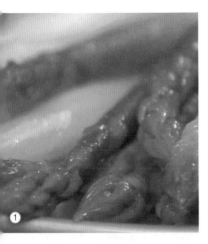

1

2

1. 白芦笋是餐桌上的美味

2. 用专门的拖拉机翻动土壤

3. 播下细小的芦笋种子

3

4

4. 芦笋幼苗需要一年时间生长　　5. 来年挖出根须　　6. 把根须深植大田

5

6

143

7. 每行作物上要堆积更多的土
8. 收获时节逐个检查幼茎
9. 长成的芦笋才能小心采摘
10. 收获的芦笋洗净并切去老根
11. 逐一修整外形
12. 加冰冷藏运往市场

7

8

9

10

11

12

Asparagus

It's considered a delicacy all over Europe and is usually eaten just with butter. White asparagus is an expensive dish that needs a lot of work to produce. It all starts off with the soil it's grown in. This requires a whole lot of preparation.

Asparagus plants prefer a very loose earth with plenty of fertiliser so the farmer has to prepare the ground for these delicate plants very carefully. To ensure the soil is loose enough, a special tractor is used to churn it up. Only when it's as loose as it needs to be will the farmer plant the seeds, but he's not trying to grow asparagus just yet.

The asparagus plant takes about a year to send up a shoot but this won't be harvested. The shoot is the first stage of the growing process to develop the root bundle or crown and this is what makes the best asparagus for harvest.

When the time is right, the crowns are dug up ready to be planted again. They are placed in the earth far deeper than they occur naturally. The stem that makes its way to the surface to find light becomes much longer than usual. This shoot is the delicacy that is white asparagus. It's white because it hasn't had any sun. Green asparagus has seen the sun and has a healthier vitamin and mineral content. The white variety being grown here is considered to be tastier, and the farmer earns more for growing it.

To give the shoots that extra depth, the farmer piles earth on top of each row. He then covers each row with black tarpaulin. This increases the soil temperature and speeds up the ripening process. Asparagus season is from May to June, so time is short. By raising the temperature the farmer extends production and makes the short season longer. This in turn will earn the farmer greater profit.

When the time is right, the asparagus is harvested by hand. Each shoot is checked and if it's ready it's dug up. But, each root bundle can produce several stems so the harvesters are careful to cover them up once again and leave them for a bit longer.

The delicate stems are collected into baskets and returned to the processing plant back at the farm house. Asparagus is an expensive food as it takes over two years to grow. A jeep is more than big enough to carry this day's harvest back to the farmhouse.

The next step of the process is to prepare the asparagus for the customer. This traditionally starts out with a shower. Once clean it's sent through the guillotine. Here the bottoms will be removed. Asparagus is a delicacy and the restaurants and store owners want their customers to have an experience. So they are prepared to pay for appearance as well as flavour. The freshly cut stems are rinsed once more and sent on to be graded. The price the farmer can get depends on the stems thickness, appearance and colour. The stalks are rolled over a bump and a camera records their movements. They are then graded accordingly.

The sorting system will then kick them into the right box where a worker checks each shoot again. Any damage or dirt is removed and the asparagus is then ready to be chilled.

Each basket is immersed in cold water and covered in ice. This stops the plant ripening any further and keeps them fresh until they can be sold. The cases are then put into cold storage and left there until the farmer can deliver the order to his customers.

It can be used as a garnish, or cooked with butter. The popular delicacy that is white asparagus.

甜辣酱

扫描二维码，观看中文视频。

泰国，是个以寺庙和烹饪中使用大量辣椒而著名的国家。在英国，越来越受欢迎的一种调料，是泰国甜辣酱。这个辣椒酱深受人们喜爱的秘密是它的辣劲。

辣椒的辣来自一种化学物质，叫作辣椒素。和一般认识相反，它在种子里含量并不是最高。辣椒素浓度最大的地方，是种子附着的薄膜。如果你足够疯狂敢咬一口新鲜的辣椒，辣椒素就被释放出来。它与你舌头上的味觉和疼痛感受细胞产生反应。发出紧急的电信号，告诉你的大脑应该停止吃进嘴里的一切东西。同时内啡肽也被释放出来，所以身体同时经历着痛苦和快乐，这也许能解释辣椒为什么大受欢迎。

有些人是如此渴望这种"痛并快乐"的糅合，他们甚至参加吃辣椒比赛。生食辣椒的世界纪录是在1分钟内吃15个加拉皮诺辣椒，但如果旁边没有医疗救护，我们不建议你这样做。

要做甜辣椒酱，首先你需要辣椒。每一年的收获季节，泰国乡村都会变成鲜艳的红色。辣椒从种子长到果实需要75天，所以一年可以收获几季。

农民看到辣椒变成鲜红色时，就知道它们成熟了。这鲜红色会警告动物它们很辣，吃起来危险，但对于农民来说这红色意味着辣椒成熟可以采摘了。这是项劳动密集型的工作，但在泰国有很多人手，因为人力比昂贵的农用设备更便宜。

只有红色辣椒用来做酱，因此未成熟的绿色辣椒被留在植株上。采摘完毕后，红辣椒就被摆开分类，以确保质量良好。将不合格的绿色辣椒拣出，把虫子除掉。然后收获的辣椒装进口袋，用卡车送到工厂。

泰式烹饪在世界各地流行，甜辣椒酱经常和泰国鱼饼一起吃，制作从辣椒开始。首先它们被研磨成糊状。不夸张地说，即使你吃了很小的量，也会泪流满面。这是纯粹的辣椒，虽然还有其他原料，但它是辣椒酱的关键成分。为了做出140瓶的辣椒酱，厨师们需要多至17千克的辣椒糊糊。没有辣椒怎么能称为辣椒酱呢？其他成分包括盐，还有大蒜。这批辣椒酱需要4千克的蒜泥。

第一种放入锅内的成分是水和醋。这是辣椒酱的基础。厨师们也往里加入面粉，这有助于使酱汁变稠。但除了红辣椒以外，加进这些混合物的最多成分是糖，约占总配料30%。糖使甜辣椒酱有了甜味。所有成分被不断地搅拌，同时又添加更多的糖。所有的糖溶解之后，就到加辣椒的时候了。他们用一个抹刀，在整个混合物中均匀分开，然后加大蒜。

现在可以开始烹调了。需要在85摄氏度下花30分钟把一切混合起来。工作人员密切观察温度和搅拌过程。没有散开的辣椒酱味道差得多。

你知道吗？

多西特那加辣椒的辣度为923000史高维尔辣度单位，是墨西哥辣椒的11535倍。处理它时要戴手套！

当烹调完成后，辣椒酱还相当稀。所以把它转移到新锅里。然后用天花板上风扇冷却。流过辣椒酱的空气使它冷却和变稠，成为我们熟悉和喜爱的那种传统质地。然后装瓶。送到客户之前唯一剩下需要做的就是让老板来品尝一下。就这样，辣椒从植物的果实变成了一瓶美味的泰国甜辣酱。

1. 泰国餐饮因辣椒而闻名
2. 辣椒素最浓的地方是种子附着的薄膜
3. 吃辣椒比赛有医护人员在场
4. 辣椒从种子到结果约 75 天
5. 泰国采摘辣椒是人力密集型劳动
6. 把洗净的辣椒绞碎

7. 放入水、醋、盐、面粉和约占总量
   30% 的糖
8. 辅料搅匀后加入绞碎的辣椒
9. 混合物搅拌均匀后加大蒜泥
10. 在 85 摄氏度下熬制 30 分钟
11. 辣椒酱转移到新锅
12. 用风扇冷却后装瓶

Sweet Chilli Sauce

Thailand, a country famous for its temples and its cooking, which contain a lot of, yes, chillis. One key ingredient that's growing in popularity in the UK is sweet Thai chilli sauce.

The secret to this sauce is the chilli-kick we love so much. The chilli's power comes from a chemical called capsaicin. But contrary to popular belief, it isn't strongest in the seeds. Its greatest concentration is in the membrane that the seeds are attached to. If you're crazy enough to bite a fresh chilli, the capsaicin is released. It reacts with taste and pain receptors on your tongue. They send urgent electric signals telling your brain to stop eating whatever is in your mouth. At the same time endorphins are released, so the body experiences pain and pleasure which may explain why they are so popular.

Some people crave this pain-pleasure cocktail so much they enter chilli eating competitions. The world record for eating raw Jalapenos is 15 in one minute but we don't recommend you try this without medical assistance standing by.

To make sweet chilli sauce, first you need chillis. Every year during the harvest the Thai countryside turns a bright red. From seed to fruit takes about 75 days so several harvests can be made each year.

The farmers know the chillis are ripe when they turn bright red. This may warn animals they're spicy and dangerous to eat, but to the farmers this means they're ripe and ready to pick. The work is very labor intensive but there are plenty of hands to help with the harvest in Thailand as people power is cheaper than expensive farm equipment.

Only the red chillis are used in this sauce so the unripe green one's are left on the plants. Once they've been picked the chillis are laid out and sorted to make sure they're the quality is good enough. Any errant green chillis are removed and any bugs are thrown out as well. The harvest is then bagged up to be collected by truck and taken to the factory.

Thai cooking is popular all over the world and the process of making this sweet spicy sauce to accompany some Thai fishcakes starts with the chillis. First they are ground into a paste. It's fair to say that if you ate the tiniest amount of this it would bring burning tears to the eyes. This is pure chilli. although there's much more to this sweet sauce than chilli, it is the key ingredient. To make just 140 bottles, the cooks need a massive 17 kilos of the ground paste. Well it wouldn't be chilli sauce without

it now, would it? Other ingredients include salt. And then there's the garlic. This batch calls for 4 kilos of mashed garlic.

The first ingredient into the pot is a mix of water and vinegar. This forms the basis of the sauce. To this, the chefs will also add flour which helps make the final sauce thicker. But the biggest addition to this mix, apart from the chillis, is sugar. Nearly 30% of the ingredients in total. That's what makes it sweet chilli sauce. Everything is continually stirred to combine the ingredients whilst still more sugar is added in. Then once all the sugar is dissolved its time for the chillis. They are added using a spatula which helps divide them up throughout the mixture. They are then followed by all that garlic.

And now they can start the cooking. It takes about 30 minutes at 85 degrees Celsius to combine everything together. The staff keep a close eye on the temperature and the mixing process. Lumpy sauce wouldn't taste nearly as good.

When the sauce is finished cooking, its still quite fluid, so it all gets transferred to a new pot. This is then taken to the nearest ceiling fan and left to cool. The air flowing over the sauce cools and thickens it, giving it the traditional texture that we know and love so much. It can then be bottled up and the only thing left to do before its sent to the customers, is for the boss to have a quick taste. From the fruit on the vine to a tasty bottle of sweet thai chilli sauce.

Did you know?

At 923,000 Scoville Heat Units, the Dorset Naga Chilli is about 11,535 times hotter than a Jalapeno. Wear gloves when handling them!

普罗旺斯香草

普罗旺斯在法国南部,这里以美丽的风景、古老的村落和长满芳草的田野而闻名。多年以来,他们开发出一种特殊的混合香草,被称为普罗旺斯香草。除了其他成分,香草含有迷迭香、紫苏和百里香,它们被干燥处理以保存气味。

在进入厨房碗柜之前,香草需要耕种。这里明媚的阳光和肥沃的土壤,为香草生长提供了最完美的环境。但在炎热的夏季月份,它们需要稍许降温,所以种植者用喷淋系统浇水。

9月中旬,田野里的迷迭香进入鼎盛时期,准备在加进百里香、茴香、马郁兰、香薄荷、紫苏和鼠尾草后,共同做成普罗旺斯香草。

收获之后,迷迭香停留在木架上干燥十天。热空气通过板条上升,香味被存留下来。然后,它被送入一台机器,把木质的茎除去。

此后它被装进大麻袋,带到一个仓库进行处理。每年有60吨的香草被送到这里。每个运来的布袋都仔细称重并做好纪录,以便追溯货物的来龙去脉。

迷迭香经过摇动的筛选机。清除污垢颗粒以外,机器的筛子网格从粗大到细小,能给香草分类。不同级别的风味有轻微的差异,其他香草也做同样处理。然后前往普罗旺斯香草搅拌机。

叉式起重车抬起箱斗,随着手柄的转动,整箱的迷迭香倾倒进去。实际上迷迭香的重量是仔细称过的,将与其他香草结合起来,达到精确的混合。

充分搅匀后,混合香草通过溜槽来到袋子里。每一袋有25千克。最后袋子被缝合起来。它们发送到世界各地,但无论最终去到哪里,都将带来一点法国南部的味道和气息。

你知道吗?

克里维斯的安妮在与亨利八世的婚礼上戴着迷迭香,因为迷迭香被认为是忠诚的象征。

1. 法国普罗旺斯很适宜香草生长
2. 普罗旺斯香草是特殊的混合香草
3. 迷迭香是普罗旺斯香草的重要成分
4. 夏天要给香草喷水降温
5. 在 9 月迷迭香生长鼎盛期收割
6. 透过板条空隙的热空气让迷迭香干燥

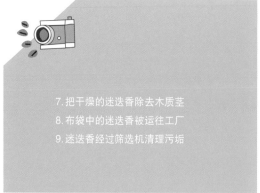

7. 把干燥的迷迭香除去木质茎
8. 布袋中的迷迭香被运往工厂
9. 迷迭香经过筛选机清理污垢

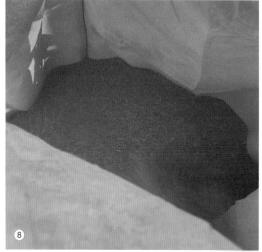

10. 机器网格由大到小
    给香草分类
11. 不同品种的香草按
    比例配置
12. 香草装袋称重封口
    后运往世界各地

Herbs De Provence

The region of Provence in Southern France is famed for it's beautiful landscapes, ancient villages and its fields full of aromatic herbs. Over the years they've developed a special blend, known simply as the Herbs de Provence. Amongst others the blend contains rosemary, basil and thyme, which will all be dried to preserve their flavours.

Before they can get into a kitchen cupboard they need to be farmed and here the sunny climate and the rich soil provide the perfect combination. But in the hot summer months they need a bit of cooling down so the growers water them with a sprinkler system.

In mid September the fields of rosemary are at their peak and it's ready to join thyme, fennel, marjoram, savory, basil and sage to make the Herb de Provence.

Once it's been harvested the rosemary is left to dry on a wooden rack for ten days. Where warm air streams up through the slats and preserves the aromas. Then it's fed nto a machine which strips away the woody stalks.

It's bagged up into large sacks and then taken to a warehouse to be processed. 60 tons of herbs are delivered here each year. Each sack that arrives is carefully weighed and recorded so that the origin of each delivery can be traced.

The rosemary is shaken through this sifting machine. As well as removing dirt particles, it sorts the herb as the machine mesh gets narrower from coarse to extra fine. The different grades can subtly affect the flavour and all of the other herbs get the same treatment. Then it's off to the Herbs De Provence mixer.

A fork lift truck lifts the crates up and with the turn of a handle the crate load of rosemary is dumped in. It's actually a carefully measured amount and will combine with the other herbs to create a precise blend.

Thoroughly mixed, it falls through a chute into a sack. Each one holds 25 kg. And finally the sacks are stitched up. They'll be sent all around the world, but wherever they end up they'll bring a little taste of the South of France.

Did you know?

Anne of Cleves wore rosemary for her wedding to Henry VIII as it was regarded as a symbol of fidelity.

厨具篇

餐具

扫描二维码，观看中文视频。

很久以来，刀子、叉子和汤匙便无处不在了，所以餐具制造商常常喜欢做些小实验。这款叉勺可能正流行起来，但直到它真正普及，似乎我们大多数人仍乐意使用一套标准餐具。但这些餐具是如何生产出来的？

现代餐具可以由不同材料制成。在这家工厂里，他们使用镀铬镍的钢。它不生锈，更重要的是，不像有些金属，它对食物的味道没有影响。

坯料从金属板切割出来，这种板材对于制造勺子和叉子都是理想材料。此时的金属对于手柄来说已经足够厚了，但功能的一端又嫌太厚，因此需要压平。做勺子的话，前端要求更宽，所以得多压几次。工人最终得到的东西看起来像个小铲子，完全不是用来喝汤的正确形状。于是铲子被切成正确的形状。然后放入这个压床里，让它变成一个凹形。

当工厂的这一端在折弯勺子，另外的工人在锻造叉子。同样，他们需要先做出小铲子，稍后压床会切掉不需要的金属。现在，坯料传递给了这个工人，他做的是叉子生产过程中最重要的一项工作。整天地坐在这里，为叉子刻出齿条。那是用来插食物的叉尖。

洗涤叉子已经足够烦琐了，但如果生产的是豪华餐具，你还必须确保弯折餐具前把压床清洗干净。任何污垢都会弄脏金属，生产出来的叉子就只能扔掉。

刚刚弯曲成型的餐具被送去进行快速抛光，磨掉尖锐的边缘。工人也对表面进行研磨，除掉能造成伤害的瑕疵。

到此为止，我们看到了叉子和勺子，那么刀呢？叉子和勺子是同一块金属制成的，刀却由几块金属做成。

这些刀的手柄有两面，它们结合在一起时形成中间的空腔，其中原因将在后面解释清楚。首先我们要有一个刀片，由不锈钢制造并成形，所以得送进炉子内预热。当温度足够高时，铁匠拿钳子把它取出来并用一个巨大压床将它们锻打成形。另一个压床从锤打后的钢片上冲压出刀片来。接着再送回炉内，使它们硬化。处置停当后，它们有了坚硬的内层，但外表肮脏。你肯定不希望用这样一把刀给吐司抹黄油，所以要把它们送到自动磨床刮洗干净。

现在是把刀身和刀柄安在一起的时候了。两半手柄已经被接起来，工人在其中填充沙子。使刀的重量在使用者手里感觉良好。接下来加入一些树脂，及时把所有部件粘在一起。摇晃作响的周日餐具是无法使用的，所以用这个奇怪的工具排列好每一把刀。把熔融金属倒进沙子里面。刚才加入的树脂会粘住沙子和金属，使手柄牢固。刀片变成一个盖子，把手柄密封起来。当然，现在一切又都变脏了，新餐具需

你知道吗？

加利福尼亚州的蒂姆·约翰斯顿保持着汤匙平衡的世界纪录。他成功地将 11 个汤匙放在脸上整整 1 分钟。

要有明亮的光泽。

一把刀如果没有锋利的刀刃，还有什么用呢？当刀刃被打磨锋利时，我们又正好赶上叉子和勺子的电镀过程。经过一次银浴，它们拥有了最终的豪华光泽，当然不同的客户有不同的口味。如果他们想要拉丝钢，餐具制造商会用一种坚韧的墨西哥草造成这种效果。草把表面光泽磨掉留下一种漂亮的纹理。然而也会再次把餐具弄脏，所以需要再来一次快速浸泡把它们清洗干净。

当你下次大量清洗餐具时，不要折弯它们的形状。留给餐具制造专家来做这些折弯的工作吧。

1. 叉勺一体的创意餐具
2. 做勺和叉把坯料前端压薄
3. 把铲形前端切成圆形
4. 压出凹形勺子
5. 制造叉子先冲压出齿条
6. 给弯曲好的叉子抛光

7.研磨除掉尖锐的边缘

8.将两半餐刀的手柄合在一起

9.刀片经锻打再冲压成型

10.刀柄空腔里加进熔融金属后组装

11.为餐刀做出锋利的刀刃

12.对叉子和勺子进行电镀

Cutlery

扫描二维码，观看英文视频。

Knives, forks and spoons have been around for a long time, so every now and then the bespoke cutlery makers like to do a little experimenting. This spork might just catch on, but until it does, it seems most of us are happy with a standard set of cutlery. But just how are these implements put together?

Modern cutlery can be made from a variety of substances. At this factory they use steel which is plated with chromium-nickel. It doesn't rust and more importantly it doesn't affect the flavour of the food which some metals can do.

Blanks are cut from a sheet of metal and this sheet is ideal for making both spoons and forks. However at this point the metal is thick enough for the handle, but too thick for the working end so it needs to be flattened. To make spoons the ends need to be wider, so the metal is pressed more often. What the worker ends up with looks like a miniature shovel, hardly the right shape for soup. So the shovel blade is cut into the right shape. Then it's put into this press which will transform it into a shapely bowl.

While the spoon bending is going on in one part of the factory, other workers are forging forks. Again, they need to shape the little shovels first, so a press will cut off any unnecessary metal. These blanks are now passed on to this worker who has the most important job in fork production. All day long he sits here, carving splines into the forks. Splines are the pointy bits you spear your food with.

Washing forks is fiddly enough, but if you're making luxury dinnerware, you also have to make sure you wash the press before you can actually do any bending. Any dirt would mark the metal and the fork would have to be thrown out.

The freshly bent cutlery is then sent off for a quick polish to remove all the sharp edges. The worker will also grind down the surfaces so there aren't any unwanted fork related injuries.

So, we've seen the forks and spoons, but what about the knives? Whereas forks and spoons are made from one piece, the knife is made out of several.

The handles for these knives have two sides, which will be joined together leaving a hollow in the middle, The reason for this will become clear later. But first, we need a blade. These are made of stainless steel and need to be shaped so they're sent to the furnace to be warmed up. Once they are hot enough the blacksmith will remove them with his tongs and hammer them into shape using a huge press. Then, using another press, he will cut out the blades from the hammered steel. These are then fed back into the furnace so they can be hardened. When they emerge, they're tough inside, but filthy outside. You certainly wouldn't want to use one to butter your toast with, so they're sent to a automated grinder to be scraped clean.

Now it's time to put the blade and handle together. The two sides of the handle have been joined and the worker will now fill this with sand. This gives the knife a good weight in the user's hand. Next comes a quick pinch of resin, and now, it's time to glue the whole lot together. Wonky Sunday silverware just wouldn't do, so this bizarre contraption helps him align each knife. Molten metal is poured in with the sand. The resin he added creates a bond between the two, making the handle solid. The blade then becomes a lid to seal the handle tight. Of course now everything is filthy again so the new cutlery needs a good shine.

And what use would a knife be with out a sharp edge? While the blades are getting sharpened, we can catch up once more with our forks and spoons which are now getting a plating. They're given a bath in silver which gives them their luxurious final sheen, but of course different customers have different tastes. If they want a brushed steel effect, the cutlery makers use a tough Mexican grass to create this effect. The grass grinds away the shine, but leaves a nice textured finish. However, it's also left the cutlery rather dirty again, so there's just time for one more quick dip to clean them off.

So the next time you're slaving over a huge bowl of washing up, don't get bent out of shape. Leave the bending to the cutlery making experts.

Did you know?

The World record for balancing spoons on your face is held by Tim Johnston from California. He successfully supported 11 spoons on his face for a full minute.

特百惠储物盒

这是一个特百惠聚会。英国的第一次特百惠聚会是在 1960 年，它引发了全国各地咖啡早茶会的热潮。特百惠是带来厨房存储革命性变化的技术奇迹。但即使这样的尖端产品，在存储有味道的奶酪时也遇到了麻烦。

今天的设计团队有很多灵感可采用。自从 60 年前在美国创立，特百惠公司经历了很长的发展道路。

现代计算机绘图在当今图形设计和完善过程中扮演着重要角色。不像那些老款式，对任何产品都装进简单容器。今天特百惠的设计预先考虑到精确的使用目的。要存储又软又熟的卡芒贝尔奶酪，设计人员必须考虑很多问题。第一是外观，奶酪饼是曲线的，所以设计师要让盒子模仿这种弧度。在商店出售时，就会和它的用途相关联。

特百惠盒子用高质量的塑料颗粒制成，能够用于食品储存。塑料颗粒被泵入机器，加热到熔点。当它们变成液体后，高压将熔化的塑料注入模具，制造出新的奶酪盒。每个机器内部，是新的奶酪盒盖或盒底的模具。当机器关闭和密封时，内部压力迫使液态塑料注入模具中的空间。塑料固化后，盒子就可以从机器中取出，准备装上新鲜的卡芒贝尔奶酪或成熟的戈贡佐拉奶酪了。

底座以类似的方式制成，根据设计者的方案使用不同颜色的塑料。你可能已经注意到，在盖子顶部仍然有一个大洞。当储存奶酪时有两大问题。一是冷凝，奶酪冷却时释放水分，它会凝结在容器内，这是不好的。设计师已经尝试了各种不同的技术，包括绒毛材料，让湿气散发。但真正的问题不只是让水分消散，还要让奶酪气味长久保持。他们最终想出了一种创新设计，带一层膜的盒子。这种膜允许较小的水分子跑掉，同时保留较大的奶酪气味分子在盒子中。

回到生产车间，我们可以看到这种盒子膜正在生产，但不像盖和底座那样采用注塑的方式。预先切好的膜放置到模具里，然后把塑料喷上去，两个部分结合而形成完好的过滤器。

生产中的最后一步是测试。特百惠盒子由塑料制成，正如我们所看到它能够热加工。然而一旦完成密闭设计，厂家就不希望再改变形状了。会让特百惠容器变热的一个地方是洗碗机，所以在测试实验室，工作人员从每个生产批次中取一个样品洗上多次，以确保它在清洗过程中不会变形。无论气味过滤器设计得如何高明，任何变形都会让奶酪的气味逃逸。

一旦确定设计方案可行，产品就会被组装起来，然后发送到世界各地的特百惠聚会中。

靠着智慧的膜，它允许冷凝水散发，但却不让芳醇的，浓郁的卡芒贝尔奶酪气味失散。你可以看到为什么特百惠厨房产品长盛不衰，它的营销派对也依然畅旺。

你知道吗？

每 2.5 秒钟，地球上某个地方就会有一次特百惠聚会。

1.一个特百惠聚会

2.特百惠带来厨房储存革命

3.按照使用目的进行设计

4.盒子模仿卡芒贝尔奶酪弧度

5.熔化的塑料喷进模具里制成盒盖

6.新奶酪盒盖的制作模具

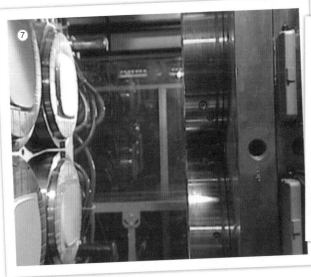

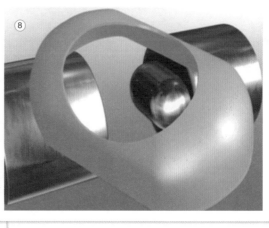

7.底座使用不同颜色的塑料

8.盖子顶部仍然有一个大洞

9.制作顶部先把薄膜贴在模具上

10.外层塑料喷上后打开模具

11.用热水浸泡检查盒子是否变形

12.带膜小孔排气并保留奶酪味道

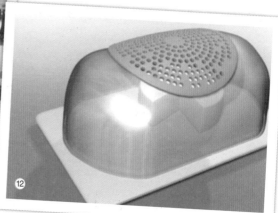

Tupperware

The Tupperware party. The first one in the UK was in 1960 and it started a tidal wave of copy cat coffee mornings around the country. Tupperware was the technological miracle set to revolutionise kitchen storage. But even that miracle product struggles when trying to store smelly cheese.

Design teams today have a lot of inspiration to draw on. Originating in the US over 60 years ago, Tupperware has developed a long way since its invention.

Modern computer graphics play a big part in coming up with and refining any new designs. Unlike old models which were simple containers for any product, Tupperware today is designed with very specific end uses in mind. To store that soft ripe camembert for example the designers have to consider many issues. The first is the appearance. Cheese wheels have curves so the designers will try and emulate that in the box. This will associate the box with its intended purpose when it's being sold in the stores.

Tupperware is made from a high-quality granulated plastic that can be used with food products. The plastic grains are pumped into a machine which will heat them up to melting point. Once they are liquid the high pressure will force the molten plastic into the mould to create the new cheese box. Inside each machine is a template for the new lid or the base of the cheese box. When the machine is closed up and is sealed, the pressure inside forces the liquid plastic into the empty space. When the plastic solidifies, it can be removed from the machine ready to cope with a fresh camembert or some ripe gorgonzola.

The bases are made in a similar way using a different coloured plastic depending on the designer's plans. Now, you may have noticed that there's an enormous hole in the top of the lid still. When storing cheese there are two big problems. 1 is the condensation. As the cheese cools it releases moisture which condenses inside your container. This isn't good. The designers have tried a variety of different techniques including nappy material to allow the moisture to be released. But the real problem is not just letting the moisture out…But also keeping the cheese smell in. The innovative design they eventually came up with for this box is a membrane which allows the smaller water molecules to escape whilst trapping the larger cheesy smelling odour molecules inside.

So back at the production plant we can see this being made, but it's not injection moulded in the same way as the lid and base. Pre-cut pieces of membrane are placed into the moulds and the plastic is sprayed onto it where the two parts combine to create the finished filter.

The final part of the production process is testing. Tupperware is made of plastic and as we have seen it can be manipulated through heat. However, once an airtight design is made, the manufacturers don't want it to change shape any more. The one place your Tupperware containers will get hot is in the dishwasher, so in the testing lab, the staff wash a sample from each batch several times to make sure it won't get warped in the wash. Any change in shape and the cheesy smell would escape no matter how well-designed the odour filter was.

Once everyone is certain that the design works, the whole unit can be assembled and then sent out to Tupperware parties all over the world.

With its clever membrane which allows condensation to escape but not the fresh odour of ripe, smelly camembert, you can see why the party is still going strong for this long-standing kitchen storage product.

Did you know?

A Tupperware party is held somewhere on Earth every 2.5 seconds.

扫描二维码，观看中文视频。

保鲜膜

无论是你第二份甜点拿过了头，还是做得太多一次吃不掉，保鲜膜都能帮上忙。保鲜膜用来保持食物新鲜，以其能够粘在附着物表面的能力而著称，它常被认为是理所当然的厨房必备品，但这种为食物保鲜的神奇塑料层是怎么生产出来的呢？

保鲜膜是用聚乙烯做成的，因为它能在食用产品中安全使用。这种塑料从原油中制造出来，先提炼成煤油。煤油被过滤和处理后就会得到固体形式的聚乙烯。颗粒状的原料通过这样的罐车运到生产工厂。大约50吨的颗粒能生产14千米长的保鲜膜。货车每周到达一次以保持材料充足。

保鲜膜就在这台机器中生产。颗粒聚乙烯被不停送入一个旋转料斗。此后被完全加热到200～300摄氏度直至熔化。两个大空气压缩机连接到料斗。它们向液体塑料鼓风吹成一个大泡泡在顶部出现。机械不断旋转，所以泡泡表面的塑料材料能够均匀分布，消除任何薄弱点。气泡还由传感器监测保证塑料膜厚度均匀。

在上面一层，这个10米高的保鲜膜泡泡被仔细压平，形成双层的膜片。卷绕到一个大辊子上准备进入后面步骤。但在我们往下讲述之前有个谜团仍需解决。保鲜膜是怎样有吸附性的？有些人认为是静电，但实际上是一种胶添加到原料颗粒里，作用像蜂蜜，为保鲜膜提供了有用的黏合性能。

现在将巨大的双层保鲜黏性膜送入这台机器，在此处被分开成单层，剪裁到合适大小。两个大卷变成10个小卷，更适合在家庭厨房中使用。

为了确保薄膜能够经得住擦划，要对新卷的样品批次进行厚度测试。但这并非保鲜膜必须通过的唯一测试。这个工厂拥有自己的实验室，随机样本被送来进行测评。保鲜膜除了黏性之外必须结实，这个测试就是测量强度的。耐久性也很重要。要实现气密性封闭，保鲜膜必须能拉伸而不会断开。这项检验测量它的弹性，保鲜膜必须能被拉伸到其正常长度的7倍，才算通过测试。

当一系列测试完成后，保鲜膜就可以包装起来卖给顾客了。使用普通的透明胶带，操作员将一卷新的保鲜膜粘到包装机上让它开始运转。

这家工厂每天能生产1500千米长的新保鲜膜，每卷这样的保鲜膜有9000米长，足够做出180筒成品。每个单筒含50米长的保鲜膜。

包装工序一刻不停。新做出保鲜膜筒被送到传送带上运到工厂的另一处准备包装。同时这台机器折叠并备好装保鲜膜的盒子。盒子和保鲜膜筒被安放到一起，一个转动的轮子将它们封装起来，准备发送到商店。这就是保鲜膜背后的故事，这个现代创新覆盖了我们的厨房，感谢它和我们"黏"在一起。

你知道吗？

使用保鲜膜和吹风机，你就能在窗户上密封一层万能的薄膜，以此来制作简易的双层玻璃。

1. 聚乙烯原料运到工厂
2. 生产保鲜膜的聚乙烯颗粒
3. 保鲜膜在这台机器中生产
4. 液体塑料被鼓风吹成大泡
5. 高达 10 米的保鲜膜泡
6. 泡泡被压平形成双层膜片

7.变成单层的大卷分成 4 个小卷

8.测量保鲜膜的厚度

9.对保鲜膜进行检测

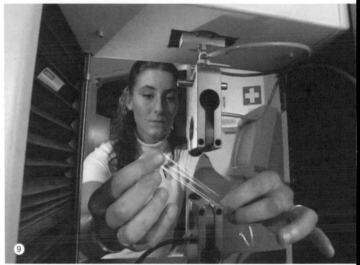

10.保鲜膜必须能拉伸 7 倍才合格

11.卷成适合家庭厨房用的小卷

12.保鲜膜进行包装

Cling Film

Whether you've gone for a second helping of dessert and overdone it, or you've just cooked too much to be eaten in one go, cling film can save your day. Used to keep food fresh and famous for its sticky ability to hang on to the surfaces it's plastered onto, cling-film is a kitchen essential that is often taken for granted. But how do they make this miracle plastic coating that keeps food fresh?

Cling film is made out of polyethylene because it's safe to use with food products. This plastic is made from crude oil, refined into kerosene. When kerosene is filtered and processed the result is polyethylene in a solid form. The granulated raw material arrives at the production plant in large tankers like this one. It takes about 50 tons of granules to make 14 kilometres of cling film. Lorries arrive once a week to keep the silos topped up.

And this is the machine that makes it happen. The granulated polyethylene is fed continually into a rotating hopper. It's then heated to anywhere between 2 and 3 hundred degrees Celsius until melted. Two large air compressors are connected to the hopper. They blast air through the liquid plastic causing it to blow up like a large bubble which emerges from the top. The machinery keeps spinning so that the plastic material is evenly distributed around the bubble surface, eliminating any weak spots. The bubble is also monitored by sensors which ensure the plastic skin has a uniform thickness.

On the floor above, the 10 metre high bubble of cling film is carefully flattened to form a double-layered sheet. This is wound onto a large roller ready for the next step. But before we go on, there's one mystery still to be solved. How does cling film cling? Some people think its static, but it's actually a kind of glue. Added to the raw granules, it acts like honey, giving the film its useful adhesive quality.

The enormous double sheet of sticky film is now fed into this machine where it's divided into separate sheets and cut down to size. Two enormous rolls become 10 smaller ones, far more suitable for use in a domestic kitchen.

To be sure the film is up to scratch, thickness tests are run on sample batches of the new rolls. But these aren't the only tests the cling film must pass. The factory has its own laboratory where random samples are sent to be assessed. Cling film must be strong as well as sticky, and this test measures toughness. Durability is also important. To create an airtight seal cling film has to stretch without breaking. This test measures elasticity and the plastic must reach 7 times its normal length to pass.

When the array of tests have been completed the cling film is ready to be packaged up for the customer. Using ordinary sellotape the operator will stick a new roll of cling film to the packaging machine and set it to work. Every day this factory can produce 1,500 kilometres of fresh film and each one of these 9,000 metre long rolls contains enough film for 180 tubes. Each tube alone contains 50 metres of film.

And the packaging process runs non-stop. As the fresh tubes emerge they're passed on to a conveyor belt which carries them to a separate part of the factory ready for packaging. Meanwhile this machine is folding and preparing the boxes for the rolls. Boxes and film are brought together and a spinning wheel closes and seals them up ready to be sent out to the shops. So that's the story of cling film unravelled, the modern innovation that's got us covered in the kitchen, thanks for sticking with us.

Did you know?

Using cling film and a hair-dryer, you can make basic double glazing by sealing a layer of the versatile film over your windows.

耐热玻璃

我们的最后一个惊人故事要从爆裂演示开始，仔细观察酒杯在喷灯下炸开……但咖啡壶却能在灼热的火焰下存留下来，这是为什么？这种玻璃耐受力惊人的秘密，全都从一家法国工厂开始。

当温度改变太快时，普通玻璃会破碎。但是硼硅酸盐玻璃或称耐热玻璃能承受快速升温或冷却的热冲击，这一切始于一堆沙子。硼硅酸盐玻璃非常适合制作实验室的烧杯和耐热厨具，能从冷柜放入烤箱再端上餐桌。

这些沙子和你在海滩上看到的沙子一样，但除去了各种垃圾和贝壳。除了沙子，科学家还将添加化学品混合物，包括氧化钠、氧化硼和回收的玻璃。

将原料放入这个炉子，燃烧温度达 1550 摄氏度左右。燃烧的火焰如此明亮，不戴滤镜就无法直视。炉内有 6 个火口和 3 个分开的腔室。

沙子和化学品混合物被不断加进第一腔室，在其中被高热熔化。随着越来越多的材料被倒入，熔化的混合物被迫通过窄孔流入第二腔室。在第二腔室中，泵进的氯气通过熔化的液体。气体吸收杂质并把它们带到表面。这使得表面层当时无法使用，但不会被浪费。表面层被移送到炉外并投入水中冷却。

这些废料玻璃，就是生产过程开始时加入沙子的回收玻璃。添加这种玻璃有助于加快熔化过程。杂质将在第二腔室再次被除掉。到达第三腔室的纯液态玻璃被均分，准备做成试管和咖啡壶。工程师们定时采取样本，检查机器是否每次取走适当分量。

对于耐热玻璃，原料必须尽可能纯净。氯气除掉了很多杂质，但工厂也充满了灰尘和污染。这些人穿上防护服以利于做些"家务活"，在这里是做工厂的工作。无尘能确保高品质的玻璃，为生产诸如咖啡壶做好准备。

因为玻璃太热，手工无法塑形，由机器进行工作。生产科学实验用的烧瓶或法压壶过程也十分相似。熔化的玻璃被放入模具，上方有气流喷射，下方有火焰加热，使玻璃延展。这将确保玻璃获得必要的强度。拉长的熔融玻璃泡被包进一个新模具中，让器皿最终成型。当玻璃冷却后，检验员从每批次中抽查样品，但它们仍有粗糙的边缘需要处理。

产品装入下一台机器中，让它们经过更多的火焰灼烧。高温会熔化不规则的毛刺，形成平滑的边缘。如果你打算用这个壶倒咖啡，它还需要一个壶嘴。边缘被加热到约 900 摄氏度，然后把凹口按压出来。

但如果你喝咖啡，那么还需要一块蛋糕。要制作蛋糕盘，熔融的玻璃需要压，而不是吹。这台机器不停工作，每天能生产 6 万个蛋糕盘。

然而暴露在极高温度下，使得这台机械需要定期维护。但是生产线不能关闭进行清洗，

你知道吗？

史前人类使用一种自然形成的火山玻璃，叫作黑曜石，来制作箭头和刀刃。

必须在操作不断进行中把弄脏的模具取走。持续地加热让模具变脏。上一个蛋糕盘携带的污垢会转移到那个模具里生产的下一个盘子上，如此反复。每个模具每16小时必须抛光除去弄脏的表层。然后由这位工程师把模具装回机器里，小心手指不被夹住。这个模具能回去继续生产新的耐温盘子，供烹调使用了。

当盘子制作出来，它们看上去像是成品了，但和咖啡壶一样，它们也需要收尾工作。盘子在火焰下旋转，封住小孔，平整瑕疵，它们会影响光洁度。透明，是耐热玻璃的好处之一，厨师可以看到菜已经做好。

正是这个了不起的玻璃制造技术，给众多的家庭厨房和专业厨房提供了极好烹饪器具——对它高温加热时也不会破裂。

1. 烈焰下酒杯炸开而咖啡壶无恙
2. 制作耐热玻璃要添加化学品
3. 原料放入1550摄氏度炉中
4. 炉内有6个火口和3个腔室
5. 工程师提取样本进行检查
6. 穿上防护服确保无尘操作

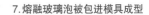

7.熔融玻璃泡被包进模具成型

8.高温熔化毛刺使边缘平滑

9.熔融态玻璃压制成蛋糕盘

10.每个模具按时抛光清除污垢

11.盘子在火焰下封住小孔和瑕疵

12.对成品盘子进行抽检

扫描二维码，观看英文视频。

Pyrex

Our final shattering story starts with a cracking demonstration … watch as the wine glass explodes under the blow-torch, yet the coffee pot survives the searing flames, but why is this? The secret to this amazingly tough glass all starts out here at a factory in France.

Ordinary glass shatters when its temperature is changed too quickly. But borosilicate or heat resistant glass can withstand the thermal shockwave of rapid heating or cooling. And it all starts with a pile of sand. Borosilicate is excellent for making laboratory beakers, and ovenproof kitchen ware that can go from the freezer, to the oven, to the table.

This sand is the same you would find on a beach, but with all the litter and sea shells removed. As well as sand, scientists will add a mixture of chemicals including sodium oxide, boric oxide and recycled glass.

The raw material is fed into this furnace which burns at around 1,550 degrees Celsius. The flames burn so brightly they can't be seen without a filter. Inside the furnace there are 6 burners and 3 separate chambers.

The sand and chemical mixture is continually fed into the first chamber where it's melted by the intense heat. As more fresh material is poured in, the molten mixture is forced to flow into the 2nd chamber through narrow vents. In the 2nd chamber, chlorine gas is pumped through the molten liquid. The gas absorbs impurities and carries them to the surface. This makes the surface layer un-usable for now. But it isn't wasted. This surface layer is diverted out of the furnace and plunged into water to cool it down.

This waste-glass is the recycled glass that's taken back and added to the sand at the start of the process. Adding this glass helps speed up the melting process. The impurities will be removed once again in the second chamber. The pure liquid glass that reaches the third chamber is divided up ready to be turned into test tubes, and coffee pots. Engineers take regular samples to check the machine is taking the right amount each time.

For heat resistant glass, the raw material must be as pure as possible. The chlorine gas removed many of the impurities, but the factory is also full of dust and dirt. These men are putting on protective clothing so that they can do some housework … or in this case, factory-work. Keeping it free of dust ensures high quality glass, ready to produce coffee jugs for example.

Machines will do the work as the glass is far too hot for human hands to shape. To produce scientific flasks or cafetieres, the process is quite similar. Molten glass is placed into moulds and blasted with a jet of air from above. This stretches the glass while flames burn it from below. This will ensure the glass acquires the necessary strength. The lengthening bulbs of molten glass are encased in a new mould which gives the jug its final shape. When they've cooled down, inspectors can check samples from each batch, but they still have rough edges which need to be dealt with.

They're fitted into the next machine which passes them over more flames. The heat melts away any irregularities to form smooth edges. Now, if you're going to pour coffee from this jug, it needs a lip. The rim is heated to around 900 degrees Celsius and the indentation pressed into place.

But if you're having coffee, then you will also need a slice of cake. To make a cake dish, the molten material is pressed, not blown.

Working continually, this machine can produce 60,000 cake dishes everyday. However, exposure to all that extreme heat means the machinery needs regular maintenance. But the production line can't be closed down for cleaning. The dirty moulds have to be removed whilst the press continues operating. Continual heating tarnishes the moulds. The dirt from each previous dish is then transferred to the next dish produced in that mould and so on. Each mould has to be polished to remove this dirty layer every 16 hours. Its then up to this engineer to return the mould to the machine without getting his fingers trapped. It can get back to work making fresh flame proof dishes for convenient cooking.

When the dishes emerge, they look like the finished article, but like the coffee jugs, they need finishing touches. The dishes are spun under a flame to seal any tiny holes or flatten any flaws that may spoil a smooth finish. Being transparent, one of the benefits of heat resistant glass is that the cook can see when the dish is done.

It's this great glass-making technology that's providing domestic and professional kitchens with cool cookware that won't crack when the heat is on.

Did you know?

Prehistoric man used a naturally—occurring, volcanic glass called obsidian. It was used to make arrowheads and blades for knives.

胡椒研磨器

扫描二维码，观看中文视频。

说说美味的现磨黑胡椒。如果没有它，胡椒牛排就不会有现在的味道。也许你知道后会感到惊讶：最近几年，这个调味料之王的销量在英国甚至超过了盐。

胡椒不同寻常的味道来自一种叫黑椒素的化学物质。它的香味可以通过加热或挤压提取出来，而胡椒研磨器是做这件事的理想工具。

这些研磨器是用山毛榉木做的，它是一种需要先干燥的硬木。但出乎人意料的是，在干燥之前要给木材喷水。这种统一的湿度能确保每片木材的干燥方式和干燥速度相同。大型通风机将水气抽出，持续工作大约 6 周时间，当木头的湿度经测量只有 8% 的时候，就可以拿去加工成胡椒研磨器了。

生产过程的第一个步骤，是把木板切割成小块。长度取决于生产的研磨器大小。这些木块将生产出约 18 厘米高的胡椒研磨器。被切割成小块的山毛榉木被送入另一个用来磨木材、而非磨胡椒的机器。在这里木块会变成人们熟悉的胡椒研磨器形状。

磨床臂在旋转的木块上来回移动，按设定的位置抬起或降低。磨床臂下的木块旋转得非常快，任何和磨床臂接触到的木材都会被磨掉。成型后的木头会经过一个砂光机被打磨平滑。现在看起来，好像人们熟悉的胡椒研磨器外形已经完成了，但接下来还有很多工作要做。

要用钻头把内部钻空。这是容纳新鲜胡椒粒和研磨机械的地方。已做成的研磨器主体排成队列，准备做一些外观修饰。

和研磨器主体一样盖子也是用山毛榉木做的。盖子就是研磨器拧动的部分，拧动将胡椒粒粉碎撒在你的比萨或新鲜意式面条上，盖子也用类似的生产过程，从一根长木料上切割下来。当木材向前推进时，磨床臂移近木头雕出盖子的形状。钻头打出螺丝孔，最后磨床臂把做好的盖子与木材分离然后开始加工下一个盖子。

与此同时，胡椒研磨器的主体部分已稍加整容。它们被放到这个传送带上送进喷漆房。首先涂一层薄薄的染料，看起来颜色更深更重。然后它们被喷上一层清漆，在整个使用寿命中提供保护。在下一道工序之前，相邻房间内漫长的蛇形旅程让清漆有时间干燥下来。

也许胡椒研磨器最有用的部分，就是研磨机械本身了。现代胡椒研磨器使用精密的瓷齿，研磨的粗细取决于不同的设置。通过扭转盖子，传输杆将旋转运动向下传送到研磨器。陶瓷做的脊或齿抓住胡椒，压碎它们并释放其中的胡椒碱。

用陶瓷材料制作研磨齿，因为它很坚韧，能长时间使用。然而，这种陶瓷不像是用来做盘子的瓷器。与陶瓷粉混合的是一种塑料黏合剂。这使得生产更容易。

在这台机器内，用 160 摄氏度高温将它熔

你知道吗？

法国汽车制造商标致，最初以其广受欢迎和成功的胡椒螺纹磨床系列而闻名。

化，并注入模具中，形成研磨器的形状。成品件一个接一个推出来，但要把它们变成非常结实的陶瓷，必须在高温下焙烧。这个炉子将部件加热到1670摄氏度。当温度超过360摄氏度时，陶瓷混合物中的塑料黏合剂被烧掉。剩下的就是纯陶瓷形态。这些部件将在这里停留两天，以便很好地硬化。一旦冷却，这些陶瓷部件会非常坚硬，足以磨碎一块花岗岩。

剩下的全部工作，就是把完成的构件组装进木制的主体部分。需要用一个塑料外壳，机器把各部分安装在一起。机器需要手工放置零件，操作员的工作就是把所有的零件发送到位，使机器人能够接管。整个研磨器用螺丝拧在一起，确保配合紧密，制作到此完成。凭着把岩石碾成粉的力量，现代研磨器粉碎胡椒的本领，在家庭和餐馆到处都受到赞赏。

1. 通过研磨器提取胡椒的美味
2. 山毛榉木湿度达 8% 时切割成小块
3. 在磨床上进行加工
4. 钻头把内部钻空
5. 研磨器主体列队进入下一道工序
6. 盖子是研磨器拧动的部分

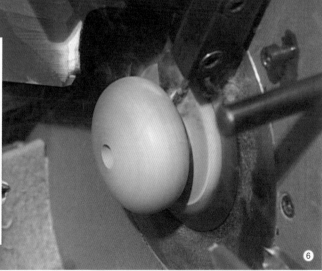

7. 胡椒研磨器喷上清漆慢慢干燥
8. 研磨机械使用精密的瓷齿
9. 传输杆带动磁齿将胡椒压碎
10. 高温下陶瓷中的塑料黏合剂被烧掉
11. 操作员将零件分配到位
12. 胡椒研磨器组装线

Pepper Mills

But first, Tasty freshly ground black pepper. No pepper steak would be the same without it. And it may surprise you to learn that in recent years, this king of condiments has even outsold salt in the UK.

Its signature taste comes from a chemical called chavicine. Its flavour can be extracted through heating or crushing, and a pepper grinder is ideal for the job.

These grinders are made out of beech, a hardwood that needs to be dried first. But surprisingly, before it's dried, the timber is sprayed with water. This uniform wetness ensures each piece dries at the same rate in the same way. Big ventilation fans draw the moisture out over about 6 weeks and when the wood measures only 8% humidity, it's ready for processing into pepper mills.

The first step of the process is to cut the planks down. The length will depend on the size being produced. These blocks will make pepper grinders around 18 centimetres high. The pre-formed blocks of beech are then fed into another machine which grinds wood instead of pepper. Here they're given their familiar shape.

As they pass back and forth over the spinning blocks, the grinder arms are raised and lowered in set positions. The blocks below are spinning so quickly that the arms grind away any of the wood they come into contact with. When they're finished, the wooden forms are passed over a sander which smoothes them. Now it may seem like the familiar shape of the grinder is complete, but there's still plenty more work to be done.

This drill hollows out the inside. This is where the fresh peppercorns and the grinding mechanism will sit. The finished bodies are then lined up ready for some cosmetic work.

As well as the bodies, the lids are also fashioned from the beech. This is the part that's twisted to crush the peppercorns over your pizza or fresh lasagne. A similar process is used to carve the lids from one long plank. As the wood is pushed forward grinding arms move in and shape the lid before a drill cuts the hole for the screw and the final arm breaks it off before starting on the next lid.

Meanwhile the grinder bodies are being given a bit of face lift. They're loaded onto this conveyor which carries them to the painting room. First they receive a fine layer of pigments giving them a richer darker colour. They're then sprayed with a coating of varnish to protect them throughout their useful lives. A long snaking journey through the next room gives them time to dry, before the next step begins.

Probably the most useful part of the pepper grinder is the grinding mechanism itself. Modern pepper mills have sophisticated ceramic teeth which grind coarse or fine depending on how they're set. By twisting the lid, the arm carries the circular motion down to the grinders. The ridges or teeth in the ceramic device draw the peppercorns into their grasp, crushing them and releasing the chavicine inside.

Ceramic material is used for the teeth because it's resilient and will last a long time. However this ceramic isn't like the china used to make dinner plates. Mixed in with the ceramic powder is a plastic binder. This makes it easier to work.

Inside this machine, it's melted to 160 degrees Celsius and injected into moulds to form the shape of the grinder pieces. The finished pieces are forced out one by one, but in order to turn them into the extra tough ceramic they must now be fired at searing temperatures. This oven heats the parts to 1,670 degree Celsius. As the temperature passes 360 degrees, the plastic binder is burnt away leaving pure ceramic shapes. These pieces will remain in here for two days, hardening nicely. Once cool, the pieces are so tough they can even grind a piece of granite.

All the remains is to assemble the pieces ready to be installed in the wooden bodies. A plastic housing is used and the parts are machined together. The machinery needs to be loaded by hand so this operator's job is to put all of the pieces into place so the robots can take over. The whole unit is screwed together, ensuring a tight fit, and the job is done. With the strength to crush pure rocks to powder, the modern grinder's pepper crushing power is appreciated in homes and restaurants everywhere.

Did you know?

French car-maker Peugeot was originally known for its popular and successful range of pepper grinders.

玻璃酒杯

先来看看玻璃葡萄酒杯。葡萄酒的风味不仅仅由葡萄或者葡萄园决定，酒杯也会对它产生影响。这个工厂每天生产 25 万个玻璃酒杯。但在认识大规模生产之前，你必须从一个原型开始考察。

你也许没有太留意酒杯的形状。其实，每个玻璃酒杯都是经过科学设计的。以增进葡萄酒特有的风味。举个例子，红葡萄酒里有一种叫单宁酸的苦涩物质。

它只有在和空气接触的时候才会变得醇厚。所以为消除一些苦涩感，酒杯必须有个宽的"肚子"，让红酒"呼吸"。你应该只倒入三分之一杯的量，然后回旋晃动葡萄酒，加速它和氧气的作用。对白葡萄酒就恰恰相反，它们有种微妙的馨香，对酒的风味至关重要。这种香味在和空气接触后就会逸散。所以你需要一个细长的酒杯，留住酒中具有的香味。

电脑模拟出三维图像，然后将酒杯原型用手工制作出来。

玻璃工匠必须不停转动吹管，否则这些炽热的白色半流体物质就会滴落到地上。这个橡木模具是按照玻璃杯设计师的精确数据定制的。玻璃被吹进来后，约需 15 秒时间定型。工人加热杯身底部，然后再加上另一小团炽热玻璃。经过切割、拉抻，然后变为杯梗。再加上最后一小团炽热玻璃，在模具里将它压成杯座。原型制作成功了，这个酒杯设计马上要投入批量生产。在工厂里，传统的手工流程让位给了高科技生产线。

这种玻璃的主要成分是石英砂，占混合物总成分的 70％ 左右。工人们放入一些苏打来降低玻璃的熔点，掺进石灰有助于稳定化合物，铝和钛则能使玻璃更坚固。这些混合物被送进熔炉，在那里，和工厂从废品站回收的碎玻璃会合。

当温度达到 1500 摄氏度时，原材料熔化并结合在一起。在这个温度下，玻璃的黏稠程度与温热的口香糖相当。36 个小时以后，它们被加热的管子送走，从一个开口挤出来，按照 240 克一份切开。玻璃仍处于白热状态，大约 1200 摄氏度。每份熔融的玻璃被压扁，一道来自上端中心的气流，将玻璃大致吹成葡萄酒杯的样子。当它们被封进一个水冷模具后，设计中的精确形状就被制作出来。

5 秒之后，模具打开。玻璃已经被冷却到 400 摄氏度，但仍然非常脆弱。每个杯子顶端都有一个临时的盖子，用来携带杯子，在工厂运输中不受损坏。

将杯身颠倒放置并用火焰加热底端，然后得到杯梗和杯座。再次浴火使杯座定型和抛光。经过这场火的洗礼，玻璃需 1 小时逐渐冷却。

用激光切掉临时的盖子，它们将会被熔化，再次进入制作流程。锋利的边缘被砂纸打磨，但对于饮用，仍然过于粗糙。于是杯沿又一次

你知道吗？

制造一瓶葡萄酒，平均要用去 600 颗葡萄。

加热到 1200 摄氏度，并在几秒钟内变得足够平滑，完全适合与嘴唇接触。

最后检查确认没有气泡和裂纹，葡萄酒杯就可以放心地打包运走了。在这家工厂，他们将传统的工艺变成了现代化的生产流程。

1. 饮用白葡萄酒或红葡萄酒要用不同的杯子
2. 红葡萄酒要加速与氧气接触
3. 白葡萄酒需要细长的酒杯留住香味
4. 制作酒杯先模拟出三维图像
5. 液态玻璃吹进橡木模具后定型
6. 小团炽热玻璃切割拉抻变为杯梗

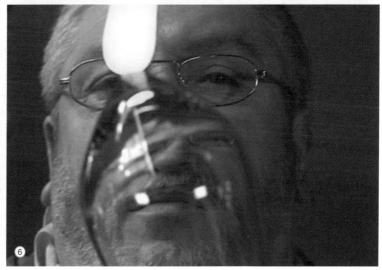

7. 再加一小团玻璃将它压成杯座
8. 去掉盖子，手工制作的原型完成
9. 黏稠的玻璃按 240 克一份挤出压扁

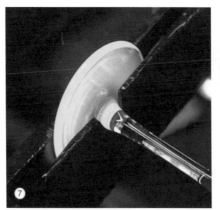

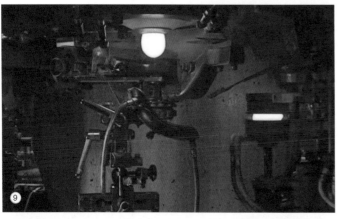

10. 模具打开后杯子顶端有临时盖子
11. 酒杯得到杯梗、杯座并再次浴火
12. 打磨后高温处理杯口

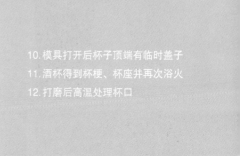

Wine Glass

But first wine glasses. It's not just the grape or vine yard which affect the flavour of the grape wine, the glass makes a difference too. This factory makes a quarter of a million of them every day, but before mass production you have to start with a prototype.

Now you may not pay enough attention to the shape but each glass is scientifically designed to enhance the flavour of the wine that it is intended for. For example there is a bitter agent in red wine called Tannin. This only mellows when it comes into contact with air. So to loose some of the harshness the glass has to have a wide bowl which can let the wine breathe. You should only pour a third of the glass and then you will be able to swirl it around to speed things up. It's the other way round for white. They have subtle aromas which are important for the flavour. These aromas drift away when they make contact with the air so you need a small narrow glass to keep the aromas where they belong.

A 3D image is modeled on a computer and then a prototype is made by hand.

The blower has got to keep the pipe moving or the white, hot, semi-fluid substance would drop to the floor. The oak mould has been made to the glass designer's exact specifications. The glass is blown in and set in about fifteen seconds. They warm up the bottom of the bowl and then add another blob of glass. That's cut, stretched and then turned to make the stem. They add a final blob and then press it into a mould to make the base. The prototype is a success and the design is ready to go into mass production. Over at the factory hand made processes give way to a high-tech production line.

The main ingredient of this glass is quartz sand, which accounts for about seventy percent of the mix. They drop in some soda to lower the glass's melting point, lime helps to stabilize the compound and aluminum and titanium make the glass tougher. The mix is lifted up to the melting furnace where it is joined by pieces of glass that the factory recycle from waste and seconds.

At 1,500 degrees Celsius the materials melt and fuse. At this temperature glass has the consistency of warm chewing gum. Thirty six hours later it's carried away in heated tubes and then squeezed through an opening and cut into ortions of two hundred and forty grams. The glass is still white hot about twelve hundred degrees Celsius. The portions are squashed flat and then a stream of air from above the centre forces them into roughly the right shape for a wine glass bowl. The exact shape of the design is created as it is enclosed in a water cooled mould.

After five seconds the mould opens. The glass has cooled down to four hundred degrees but it is still extremely fragile. Each glass has a temporary lid on top of it's bowl, which is used to carry it around the factory without it getting damaged.

The bowl is turned upside down and the bottom is heated with a torch. Then they get their stems and bases. Another blast of fire shapes and polishes the bases. After their Baptism of fire the glasses are given an hour to cool off.

A laser lops off the temporary lids. These go off to be melted and start the process again. The sharp edges are smoothed with sandpaper but are still too rough to be drunk from. The rims are then reheated to twelve hundred degrees Celsius and get smoothed out in just a few seconds; smooth enough to be touched by lips.

After a last check for air bubbles and cracks the wine glasses are ready to be safely packed and shipped. In this factory they have turned a traditional craft into a modern production process.

Did you know?

It takes an average of 600 grapes to produce a bottle of wine.

软木塞

制作一瓶上乘的葡萄酒，不只需要最优质的葡萄，还要一个好的软木塞。你可能没有意识到，但软木确实是从树上长出来的，这种树大部分生长在葡萄牙。

这些人用斧头把树皮从软木橡树上砍下来，这就是用来制造软木塞的材料。看起来像是一个很粗鲁的操作，但他们是剥树皮的专家，不会对树造成损害。需要9年的时间才能重新长出足够厚的树皮，以供再次砍伐。但由于砍伐树皮过程中采取的小心措施，每棵树在150年的寿命中可以被砍伐多次。

单单在这片树林里就有大约200万棵树，每棵能提供20千克树皮。大量的树皮被采集后，带回到这样的场院。这里有15000吨软木的树皮。工人首先烹煮树皮，使之更加柔软和坚韧。树皮在工厂里用大栈板运送。这个栈板中装有超过1300千克的树皮。树皮洗净后运回场院放置半年左右，让它从空气中吸收水分。

大约6个月后，树皮被取回室内用圆锯切成长条。长条树皮直接来到工人手中被冲压成软木塞的形状。冲压后剩下的边角余料不会被浪费，我们稍后将会看到它们的用场。随着瓶塞沿传送带行进，一个自动系统根据它们的孔隙特性进行识别并分拣。现在是用人工的时候了。这些妇女根据质量将软木塞归入不同的批次。她们必须正确操作，因为最好的软木塞每

个能卖到50便士（1英磅=100便士）但那些不太好的只有几便士。在实验室中，必须检查不同软木塞接触到液体时如何反应。有时候发霉的软木塞会让酒变质，被称为"软木塞味"。通过测试各批次样本，能够预知是否会霉变，并尽可能阻止发生。

这就是那些边角余料的最终归宿。它们被粉碎成小块，然后送到一台机器里与黏结剂混合。再将混合物加热加压，形成这些2米长的软木塞。不起眼的边角料实际上将被用于最好的酒，香槟。2米长的软木塞被切到合适的尺寸后，在一端粘上两个圆盘。直到木塞进入香槟酒瓶之前，它们不会具有人们所熟悉的形状。

在生产线的终端，最好的软木塞被贴上标签并分装成袋，准备前往葡萄园的旅程。它将被塞进酒瓶，在你身边的晚餐宴会上，用螺丝锥开启。

你知道吗？

在被开启的时候，香槟软木塞的平均飞行速度为每秒13米，接近每小时30英里。

1. 上乘葡萄酒还要有个好的软木塞
2. 树皮从软木橡树上砍下
3. 剥光树皮的大树 9 年才能复原
4. 树皮烹煮洗净在场院放置半年
5. 用圆锯将树皮切成长条
6. 在长条上冲压出软木塞

7.冲压后的边角料将再利用　　10.用边角料生产出的软木塞条

8. 按质量将软木塞归入不同档次　11.切成合适长度在顶端贴上两个圆

9.测试样品，防止酒中有软木塞味　12.贴上标签并分装成袋

Corks

扫描二维码，观看英文视频。

But first, You don't just need the best grapes to make a great bottle of wine, you also need a fine cork. You might not have realised it but cork really does grow on trees and most of those trees are here in Portugal.

These men hack the bark from the cork oak trees with hatchets. this is what will be used to make the corks. It looks like a rough process but these guys are specialist debarkers and are not harming the trees. It'll take 9 years for the bark to grow thick enough to be cut again. But with the care they're taking removing it they'll be able to harvest it many times over the 150 year life of each tree.

There are around two million trees in this forest alone and as they provide twenty kilos of bark each, huge amounts are collected and brought back to yards like this one. There are fifteen thousand tons of cork slabs here. First of all they boil the bark to make it softer and more flexible. It's moved around the factory on large pallets. This one is loaded with over thirteen hundred kilos. After it's been washed the bark goes back out into the yard for around half a year so it can absorb moisture from the air.

About 6 months later it's brought back in doors where it's cut into strips by a circular saw. The strips go straight to these workers who punch the cork shapes out. The off cuts aren't going to be wasted, we'll find out what happens to them a bit later on. As the corks file along this belt an automated system checks and sorts them according to how porous they are. Then it is time for a human touch. These women sort them into batches of different qualities. They have to get it right because the best corks can go for fifty pence each but the lesser ones cost just a few pence. In the lab they have to check how the different corks react when they come into contact with liquid. Sometimes mildew builds up on a cork turning the wine bad. That's when a bottle is referred to as being corked. By testing samples from the batches they can tell if that's likely and stop it from happening as much as possible.

This is where those off cuts have ended up. They're crushed into small pieces and then taken off to a machine where they'll be mixed with adhesive. The mixture is then heated and pressed to form these two metre corks. The humble off cuts are actually going to be used for some of the finest wine of all, champagne. After the two metre corks are cut down to size, two discs are glued on top. The champagne cords won't get their recognizable shape until they are actually put into the bottle.

At the posher end of the line, the best corks are labeled and sorted into bags, ready to make their way to the vineyards, into a bottle and on to a corkscrew at a dinner party near you.

Did you know?

The average champagne cork flies at 13 metres per second, that's nearly 30 mph.

橡木桶

在 2005 年，英国居民喝了超过十亿升葡萄酒。它们从瓶子里到达餐桌上，但真正的生涯却是从酒桶开始的。

这家传统的德国制桶商，正要制作一个约 225 升容积的大桶。制桶匠，就是做桶的人，需要 3 种长度的木料。将它们分为 4 块，用来制成所需要的木材。做桶不只是把木头锯成厚板，再粘到一起。这样的话最终只能得到一个方桶。

锯木头的另一个问题是树的微孔。当树成长时，微孔有助于携带营养物质到达树叶和树枝。但是，一旦树被砍倒，微孔就变成了洞。因此木板必须顺着木材的纹理锯断，以防止桶内的东西随时间推移而泄漏。

传统上用来做桶的材料是橡木，因为它很结实，并且释放出来的化学物质似乎能改善酒的质量和风味。

用木材做桶板，首先需要刨平，末端按定角度锯断，以便塑造酒桶的形状。要做出正确形状的桶，桶板必须中间宽两头窄。制桶师傅需要多年的经验才能用手工做到这一点。木桶大小并不完全一样，传统的葡萄酒桶直径有 56 厘米。为了获得合适的尺寸，制桶师傅必须使用 25～30 块桶板，但每块桶板都略有不同，他要不断尝试，直到实现正确的组合，这有点靠运气。

当拿到备好的桶板，他会把两块固定到金属箍上，形成添加其他桶板的结构。当最后一块桶板放进去时金属箍可以收紧了。瞧，一个完美的酒桶，当然还没完全做好。你看到桶的形状开始显现了，但还不能真正用它装酒。

首先桶板需要弯曲。制桶师傅在桶里点上小火，往桶板上刷水。热气和湿气共同让木材变得柔韧，因此可以弯曲。但并非用手来掰弯。要装上一个大虎钳慢慢收紧。这是个非常非常缓慢的操作。花 2 小时收紧后可以看到桶的形状了，虽然它还没有顶部和底部。

接下来制桶师傅要烘烤桶的内部。此时付出的劳动，将对桶中酒的味道产生极大影响。点着小火把木头烧焦。多些烘烤会使酒更具丰富的烟熏味道。但必须注意不要烧得太过分，以免毁了酒的味道。

虽然已经花了很多工夫让桶不漏水，制桶师傅此时却在桶的侧面钻了个洞，让酒可以流出来。他还需要加上顶盖和底盖，所以要在桶内切出一个边缘。制桶师傅用圆规量出桶的直径。然后，他可以把尺寸画到木板上做出底盖。再以开动他可靠的电锯，切割出一个完美的圆桶底。要想安装桶底，已经收紧的金属箍必须取掉。他插入桶底板，然后重新套上箍，把它再次收紧。为了确保完全不漏水，他把稻草插进每个木板间的缝隙里。

现在酒桶已经有型有款，但看起来仍然破旧，制桶师傅可以为它来一次整容了。过去打磨外壳常用手工完成，但要花很长时间，进度

你知道吗？

世界上最大的酒桶被称为"巨轮"，如果盛满的话可以装 765 万升酒。

缓慢并且非常艰苦。现代制桶师傅用机器旋转酒桶，帮助完成这项工作。首先把木头刨平，留心各处的大量的碎茬。然后再次用细砂纸最后打磨大桶。剩下要做的，是安装一些新钢箍。

制桶师傅在钢材上做标记，用定制的铡刀把钢材切断。接着把钢材展开让它紧贴着桶产生最大的抓力。然后铆接在一起使之强固。现在把铁箍装到桶上用锤子敲击牢牢固定到位。

最后加上的是制桶师傅的封印，好吧，是他的图章。所以下次当你在霞多丽酒中尝到橡木味道的时候，要检查一下还有没有漂浮的碎木渣。

1. 酒的真正生涯是从酒桶开始的
2. 工匠把木材锯成 3 种长度的厚木料
3. 中间宽两头窄的桶板靠经验组合
4. 先将两块固定到金属箍上再依次添加
5. 桶的形状显现但桶板需要弯曲
6. 点上小火再往桶板上刷水让木材柔韧

7. 一个大虎钳收紧后酒桶成形

8. 将桶内部木头烧焦使酒具有烟熏味

9. 在桶侧面钻洞作为酒的进出口

10. 加顶盖或底盖要先取掉金属箍

11. 机器旋转为酒桶打磨外壳

12. 安装新钢箍用锤子敲击固定

Oak Barrels

扫描二维码，观看英文视频。

In 2005 UK residents drank over a billion litres of wine. It may have reached their table in a bottle but it began life in a barrel.

At this traditional German barrel manufacturer, they're going to make one that can hold about 225 litres. The cooper, who makes the barrel, will need 3 lengths of wood. He'll split them into 4 pieces and use that to make the planks he needs. Now making a barrel isn't just about sawing wood into planks and sticking them together. You'd end up with a square barrel for a start.

Another problem with just sawing up your wood is the tree's pores. When the tree was growing, they would help carry nutrients to the leaves and branches. However, once it's cut down, the pores become just holes in the wood. The planks must therefore be cut with the grain of the wood to stop the barrel's contents leaking out over time.

Traditionally oak is used to make barrels because it was strong and the chemicals it releases seem to improve the quality and flavour of the wine.

To make staves from the planks, they're sanded down first. The ends are then sawn off at an angle to help shape them. If you want a proper barrel shape the staves must be wider in the middle and narrower towards the ends. It takes a master cooper years of experience to be able to achieve this by hand. Now, although they're not all exactly the same, traditional wine barrels have a diameter of 56 centimetres. To get this right, the cooper must use between 25 and 30 staves but as they're all slightly different, he has to keep trying until he gets the right combination which can be a little bit hit and miss.

When he's got the one's he needs, he'll attach two of them to a metal hoop. They'll form the structure to which all the others will be added. Once the last one is ready to be put in, the hoop can be tightened up and voila, one perfectly formed wine barrel, well not exactly. You can see the beginnings of a barrel shape but you couldn't really store any wine in it just yet.

First the staves need to be bent. The cooper lights a small fire in the centre of the barrel and the staves are brushed with water. This combination of the heat and the humidity makes the wood flexible so he can bend it. But it's not a case of just bending them in by hand. A large vice is attached and slowly tightened up. It's very, very slow. After about 2 hours of tightening, you can actually see the barrel shape even though it doesn't have a top or a bottom yet.

Next the cooper will toast the interior. What the cooper is doing here will have a really big impact on the flavour of the wine stored in this barrel. A small fire is lit to singe the wood. More singeing will mean a richer, smokier flavour for the wine. But he has to be careful not to burn the barrel too much. That would ruin the wine.

Having spent ages making it water tight, the cooper now drills a hole in the side so he can get the wine out. He also needs to give it a top and a bottom so a lip is cut into the inside. The cooper then measures out the diameter using a pair of compasses. He can then transfer the size to the boards he's using to make the bottom with. It's then time to turn on his trusty saw and cut out a perfect circle for the bottom of the barrel. But to fit it, the hoops that have been holding it together have had to be removed. He inserts the disc, and then replaces the hoops and tightens it all up again. Then just to be sure it really will be waterproof, he sticks straw into the gaps between each stave.

Now, he's got the barrel into shape, but it still looks a little bit shabby round the edges so the cooper can now give it a bit of a make over. Sanding down the exterior used to be done by hand, but it was long, slow and very hard work. The modern cooper is far smarter. He gets his assistant to do the work using a machine to spin the barrel for him. First he planes down the wood, watching out for enormous splinters as he goes. And the barrel is then given the once over with some fine sandpaper to finish it. All that remains is for the barrel to be fitted with some fresh steel hoops.

The cooper marks up the steel and cuts it on his custom guillotine. He then splays the steel so that it's tight against the barrel and gets the best grip possible. It's then all riveted together for strength. The hoops are now fitted to the barrel and hammered firmly into place.

And the final addition is the cooper's seal, well ok his stamp. So the next time you taste oak in your glass of Chardonnay, check there aren't any splinters still floating around.

Did you know?

The largest wine barrel in the world is called the "Giant Wheel" and could hold 7,650,000 litres of wine–if it was ever filled.

冰箱生产

我们都有过这样的经历。在深夜,你饿了,想吃零食。该去冰箱搜寻点什么。但你可曾停下来想过,这种令人惊奇的发明是如何工作的?

要制造这种方便的柜子,理所当然会需要钢材。大部分冰箱的外壳使用钢铁,在这样的工厂里制造出来。一个 160 吨的压床从钢卷上压出零件。这一过程有 36 个不同步骤,质量管理员始终严密监控。

冰箱是现代生活的必需品。据估计在英国有超过 4000 万台冰箱用来冷藏食物和饮料。冰箱平滑的外壳是怎么做出来的?相反的电荷彼此吸引,在涂漆过程中让金属零件带负电,喷漆器带正电,使油漆粘住,形成平滑的表面层。

冰箱由两个重要部分组成。我们已经看到了外壳的压制和喷涂。这种塑料材料将被用于构建冰箱内胆。大功率的炉子把塑料加热到 220 摄氏度。在柔软的熔融状态,被滚压成宽的薄板。压出的塑料板材用刀片切成适当大小,然后进行生产的下一阶段——成型。

在英国,每年有 250 多万台冰箱被丢弃,简直堆积如山。作为回应,现代立法规定,冰箱必须使用比过去更加环保的材料制造。

塑料板材被装到真空成型机里。它们在这里被吹胀,将模具推进去然后吸出空气塑料紧贴到模具上形成完美的冰箱内衬。水冷让塑料的新形状硬化任务完成。

冰箱内衬装配到钢制外壳里。它们之间的空隙充入绝缘性烃类泡沫。但在此之前两个做好的部分要安装在一起。外部钢壳提供支撑结构。塑料内衬安装在里面,包含一个非常重要的孔洞用来注入绝缘材料。两部分之间的空腔现在可以填充绝缘材料了。每个工件装上机器紧固到位。它们被连接到喷嘴上,液化绝缘材料注入钢和塑料之间的空隙。泡沫膨胀并变硬,将两个部分粘到一起,并使冰箱整体绝缘。

我们可以实时看到泡沫膨胀的过程。这种膨胀型的绝缘材料曾经是破坏臭氧层的氯氟烃。制造商在 20 世纪 90 年代中期转而使用烃类化合物直到现在。冰箱结构部分完成后,可以安装调节温度的装置了。

冰箱的工作原理是将箱内热量传走。其中的食品并非直接冷却的,秘密是使用碳氢制冷剂。这些化学品通过一系列管道把热量从内部带到外界。当热量带走后食品就逐渐冷却下来。制冷剂在非常低的温度下沸腾。食品或冰箱本身的残余热量被传递给这种液体。制冷剂沸腾然后蒸汽迅速通过管道网络把热量带走。

当蒸汽通过管道来到冰箱之外后,将热量释放。碳氢制冷剂返回储存器进行冷却,再泵回循环系统。泵或称压缩机是制冷过程的最后一部分。它让制冷剂在冰箱中循环。系统必须在密闭条件下运行,所以一切零件

你知道吗?

日本有家公司设计了一款内置网络摄像头的冰箱。用户能够登录并查看以确认回家路上要买些什么。

都要焊接紧密。

　　最后将存储东西的架子、抽屉和箱子安放停当，冰箱组装便完成了。把成品冰箱包上玻璃纸防护层，并打包准备出售。当你下次光顾冰箱的时候，应该感谢这项创造性发明，它让你的现代生活如此精彩。

1.冰箱是现代生活必需品
2.制造冰箱需要钢材做外壳
3.压床从钢卷上压出冰箱外壳

4.利用相反电荷彼此吸引使喷漆平滑
5.压出的塑料板材切成适当大小
6.塑料板材加热吹胀形成冰箱内衬

7. 冰箱外部钢壳为内衬提供支撑

8. 充入烃类泡沫使外壳内衬粘紧

9. 安装调节温度装置

10. 冰箱的工作原理

11. 压缩机让制冷剂在冰箱中循环

12. 将存储物品的架子等安放停当

⑦

⑧

⑨

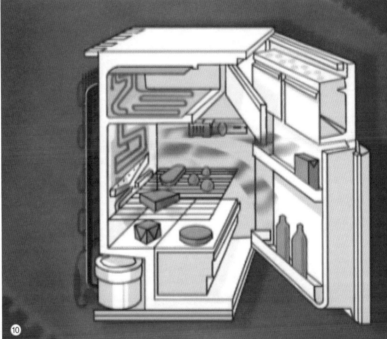

⑩

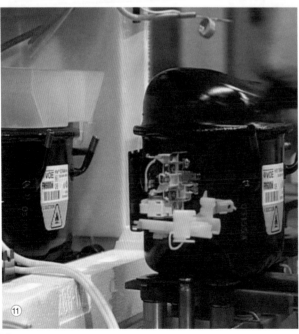

⑪

⑫

Refrigerator Production

扫描二维码，观看英文视频。

But first, we've all been there. Its late at night, you're hungry and you fancy a snack. Time to raid the fridge. But have you ever stopped to consider how this amazing invention works?

To build these handy cabinets that we take for granted you need steel. It's used to make most fridge exteriors which are fabricated at a factory like this one. A 160 ton press stamps the individual parts out of a steel roll. The process has 36 different steps and is watched by a vigilant quality control manager.

A fridge is a necessity in modern life. It's estimated that in Britain there are over 40 million fridges keeping our food and drinks chilled inside but how do they give them that smooth finish on the outside? Opposite forces attract, so in the painting process the metallic parts are negatively charged while the paint sprayer is positively charged. This makes the paint stick in a smooth layer.

A fridge is made up of two key sections. We've seen the outer shell pressed and painted. This raw plastic material will be used to build the interior. Powerful ovens heat the plastic to 220 degrees Celsius. In this soft molten form it's rolled out into wide sheets. This roll of plastic is then cut to size using a razorblade and carried on to the next stage of production where it will be shaped.

Over 2 and a half million fridges are thrown away in the UK every year which has created something of a fridge mountain. To combat this, modern legislation requires that fridges must be made more environmentally friendly than they were in the past.

The plastic sheets are loaded into a vacuum forming machine. Here they are blown up, the template is pushed inside and the air is sucked out. The plastic clings to the mould which forms a perfect new fridge interior. Water cooling hardens the plastic into its new shape, and it's ready to go.

The interior will be fitted into the steel cabinet exterior, and the space in between filled with an insulating hydrocarbon foam, but before that's done, the two finished pieces have to be put together. The outer steel shell provides the structure. The inner plastic linings are fitted inside and they include an all important hole for the insulation-injection. The cavity between the two parts can now be filled with the insulation. Each unit is loaded into the machine and held firmly in place. Here they are attached to a nozzle which injects the liquefied insulation into the spaces between the steel and plastic. The foam expands and hardens which both seals the unit together and helps insulate the fridge as a whole.

The foam can be seen expanding when we view the process in real time. This expanding-type insulation used to be made with ozone-depleting CFC's or chlorofluorocarbons. Manufacturers switched to Hydrocarbon compounds in the mid 1990's and have used them instead ever since. With the structure of the fridge complete, the technology that regulates the temperature can now be installed.

Fridges work by transferring heat away from the inside of the cabinet. The food inside isn't cooled directly, the secret is in the use of a hydrocarbon refrigerant. These chemicals carry heat away from the interior, through a series of pipes to the outside. As the heat is removed, the food is gradually chilled. Refrigerants boil at very low temperatures. Any residual heat in food or in the fridge cabinet itself is transferred to the liquid. It boils, and the steam then passes rapidly through a network of pipes; taking the heat with it. As the steam passes through the pipes and out of the fridge the heat is released. The hydrocarbon then returns to the reservoir where it cools, before being pumped back in to the system. The pump or compressor is the final part in the process. It keeps the cooling agent circulating around the cabinet. The system must be airtight to work, so everything's welded for the perfect seal.

Finally all the shelves, drawers and boxes that are used to store your shopping are added to complete the fridge. A protective layer of cellophane is wrapped round the finished unit and it's packed up ready to be sold. So the next time you're raiding the fridge, give thanks for the innovative invention that keeps modern life so chilled.

Did you know?

A Japanese company has designed a fridge with an internal web–cam. Users will be able to log on and see what to buy on the way home.

穿戴篇

雨果博斯西装

扫描二维码,观看中文视频。

每位男士的生活中都有必须穿西装的时候。也许是婚礼或葬礼,或者是一个重要的工作场合。对很多人来说买成衣就够了,但由裁缝定制的无疑最理想,剪裁得最合身。

如果能负担得起裁缝定做衣服的高昂价格,你就需要找专业的裁缝,像这两位先生一样。他们的知识和专长能帮助买家确定合适的材料和式样。

西装不只是由细条纹或粗花呢做成。有多层材料能让西装挺括和款型适当。

材料被选定后,裁缝就要去量度量顾客的尺寸,结合样式做出裁布的模板。将尺寸和规格发送到布料部门进行剪裁。模板铺在台面上,但并不是用来告诉计算机往哪里切割的。计算机处理器中已经储存了切割布料的所有信息。模板是给工人看的,让他们确认下刀的地方准确无误。

切割机完成工作后,会裁出 120 多块布料来组成服装。这个工人不是在给每片布料单独地贴价签而是给它们做标记,这样就能用正确的布料给顾客做衣服了。

有了正确的布料,每个工人便在自己岗位上完成她的那部分工作,开始把西装缝在一起。举例来说,细条纹西服的线条必须从一开始就严格对齐。裁缝用辅助工具确保这项工作做好。她小心地避免针刺伤自己,而用针固定住底布,然后在手持激光仪帮助下连接布料,接着将每块布料钉住,最后胸有成竹地把所有布料正确对齐并缝合起来。

做西装时很重要的另一个部位是肩膀。首先让胳臂到位也有助于西装造型适合每个顾客的体态。缝进垫肩有助于穿用者显得肩膀宽阔。

考虑到定制西装的花费从几百到几千英镑,最终产品的质量必须是上乘的。

每道工序都需要不断地检查。成型和缝制的过程让西服留下细微褶皱,所以要给它一次彻底熨烫以消除难看的皱痕。在专门的蒸汽设备中,整个西装被空气充胀,然后用热蒸汽熨烫。熨好的西装看起来崭新无瑕,可以上身了,但仍有一些细节有待处理。

做扣眼时先把布缝起来,这样能提供支撑使孔洞边缘不会在穿脱和拉扯中磨开。然后在缝线的中间切出孔来。

接下来是纽扣。用手缝很费事但机器让这个过程看起来非常简单。

最后对西装来说,熨烫翻领能让它看起来有崭新的样子。

一套西装需要一件上衣和一条裤子。下面来做它们。西装的膝盖部位总会受到很大的拉力,但量身定制的西服不会显出任何伸展,因为布料从内层巧妙地加固了。每一条裤腿都增加了第二层布料,以利于支撑膝盖部位。

另一个需要良好支撑的区域是腰部。这是身体转动的部位,并且大腹便便也对腰带产生

你知道吗?

在雨果博斯成为时尚界家喻户晓的名字之前,他曾为纳粹党和德军设计制服。

超负荷的压力。加强的布料被缝进去，用来支撑面料布保持裤型，让穿着者从正面和背面看上去都形象良好。牛仔裤前裆开口经常用纽扣，但西装毫无例外地用拉链。下一步就是用金属条把拉链钉上去。

每个穿过西装的人都知道，让裤子的折痕保持清晰有多难，但对于这些工人来说却很简单。工业压力机专为这项工作设计，工人把裤子放进去后启动。蒸汽和压力的共同作用将裤子熨平，为穿着者做出一个完美的折痕。

这条裤子现在可以和上装一起交付顾客了，让他穿上去具有完美的职场形象。

每个细微处都经过精心制作，这样就打造出了完美无瑕的西装，这是廉价的成衣所无法比拟的。

1. 西装是男人生活中的必需品
2. 专业裁缝能制作最合体的西装
3. 选定材料后通过电脑设计出模板
4. 根据模板对衣料进行剪裁
5. 对 120 块布料进行标记
6. 激光仪帮布料线条对齐

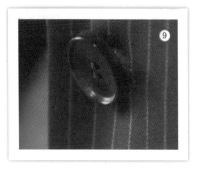

7.加进垫肩让肩膀显得宽阔

8.将西装充胀用热蒸汽熨烫

9.锁好扣眼，钉牢扣子

10.裤腿的布料从内层加固

11.腰部缝进加强布料保持裤型

12.蒸汽和压力将裤子熨出完美的折痕

Hugo Boss Suits

扫描二维码，观看英文视频。

There comes a time in every man's life he must wear a suit. Whether it's for weddings or a funeral , or that all-important job. Off the peg suits many people but surely the most desirable of all is the bespoke tailor made suit, cut to fit perfectly.

If you can afford the extravagant price tag that comes with tailor made clothing you would need to find professional tailors like these two gentlemen. Their knowledge and expertise help guide the buyer with decisions about material and style.

A suit doesn't just consist of pinstripe or tweed cloth. There are a variety of different layers to strengthen the suit and help shape it properly.

Once the materials have been selected, the tailor will take his client's measurements and combine them with the style to provide a template for cutting the cloth. The measurements and specifications are then sent down to the cloth department to be cut. The template is laid across the top, but it isn't used to tell the computer where to cut. The processor that guides the knife has all the information in its memory already. The template is for the operator to make sure the knife is cutting the cloth in the right places.

When the knife cutting machine has done its work, there are over 120 pieces left to make the suit with. This worker isn't pricing each piece individually. She's labeling them so the right pieces are used to make the clients suit with.

With the right pieces at her work station, one of many workers in this process can do her part and start putting the suit together. The alignment of the lines in a pin-stripe for example, must be perfect from the very beginning and the seamstress has a useful tool to ensure this is done properly. Using this needle carefully so as not to prick herself, she attaches the backing cloth. Then a handy laser can guide her when she attaches the suit material. The pieces are then stapled together and sewn up safe in the knowledge that everything is aligned properly.

Another element that is important when making a suit is the shoulders. Firstly they keep the arms in place, but they also helps to mould the suit to each individual customer's body shape. Padding is sewn in which also helps to emphasise the broad-shouldered appearance of the wearer.

Considering that bespoke suits can cost anything from several hundred to several thousand pounds, the quality of the final product has to be excellent. Work has to be continually checked. The process of shaping and sewing the jacket has left it a little wrinkled. So now it's given a thorough pressing to remove any unsightly creases. Using these dedicated steamers, the whole jacket is inflated with air and then pressed with hot steam. It emerges looking pristine and ready to be worn, but there are still a few details to be addressed.

To make the button holes, first you sew them up. This simply provides support so the edges of the button hole won't fray with wear and tear. The hole is then cut out of the middle of the stitching.

Next come the buttons. A fiddle to do by hand, but this machine makes the process look very simple.

And finally, as far as the jacket is concerned, the lapels are ironed in to give them that crisp new suit look.

Now, a suit needs both a jacket and a pair of trousers. They come next. Knees in suits always take a lot of strain, however when they're tailor made, the cloth doesn't show any strain and that's because the material is expertly reinforced inside. Each piece of the leg receives a second layer of material. This helps support the knees.

Another area which often needs good support because it takes a lot of wear is the waist line. This is where the body rotates, and large bellies can put strain on over-worked belt loops. Strengthening elements are sewn in which support the cloth and shape the trousers so the wearer looks good from the front and back. Button flies are often popular for jeans, but suits invariably have a zip to close them. This is added next and stapled into place with a metal band designed for the job.

Now anyone who has ever worn a suit knows how hard it is to keep those trouser creases crisp, but for these workers this job is simple. Using an industrial press designed specifically for the job, the worker will place the trousers into it and turn it on. The combination of steam and pressure will flatten the trousers and provide a perfect crease for the wearer. The trousers can now join the jacket ready for the client, giving him that all-important executive image.

Every little detail is carefully produced to create the perfect fit that's rarely found in cheaper off-the-peg garments. The tailor made suit.

Did you know?

Before Hugo Boss became a household name in fashion, he designed uniforms for the Nazi party and the German army.

扫描二维码，观看中文视频。

牛仔裤

如果没有牛仔裤，我们会怎么样？也许都在穿灯芯绒裤子。对于那些希望拥有顶级牛仔裤的人来说，这里有家工厂，专门生产名牌牛仔裤。

每条牛仔裤由 19 片不同形状的布料构成。首先要在电脑上将所有的部分绘制出来，确保用上每寸布料。图样被打印到一张巨大的纸上供剪裁。所有的部分都有编号，以确保一个口袋不会被缝到裤腿内侧。

楼下，工人把牛仔布铺好准备模板到来。根据布料不同的厚度，他们一次剪裁20～40层。这位女工对她的活计非常严肃，如果布料没有铺平，牛仔裤就会做砸。一旦全都铺好后，他们把图样放到上面，开始剪裁。

这个裁剪工看起来像电影《勇敢的心》里的临时演员。但实际上，他戴这个手套有充分的理由，为预防手指被锯断。

剪好的牛仔布被送到一个嘈杂的缝纫大厅。这里有大约 300 个工人，一天生产达 6000 条牛仔裤。他们要用掉超过 6 英里长的牛仔布和惊人的 750 英里的线，长度几乎够从伦敦一直拉到马德里。裤腿外接缝的工艺需要藏起针脚，所以他们把两个边缘折起来，从内侧缝制。牛仔裤从这一站交到下一站，每条裤子都有大约 25 个工人进行过加工。最后，他们打上铆钉。还必须用传统的方法钉好扣子。针线活完成后，一个质检员会将它快速检查一遍。

这些牛仔裤将会被做成破旧的样子。它们被充气，然后工人用砂纸进行磨损。因为是手工操作，所以每件都与众不同。当它们接着被清洗的时候，磨损处颜色会脱落使牛仔裤看上去很破旧。相不相信，这其实是整个制作过程中最昂贵的一步。工人要花很多时间在正确的位置做出线条，才能看起来像自然产生的折痕。

要让牛仔裤貌似经典的破旧款式，下一步是化学处理。工人用高锰酸钾对裤子进行漂白，这样既让染料褪色，又不会损伤布料。

更耐久的牛仔布使用喷砂打磨法。牛仔裤用钳夹固定，工作区是密封的，防止沙子到处乱飞。一个工人用 4 帕的压力，向一条牛仔裤喷上大约 1 千克的沙子。他必须很小心，不要在同一个处喷的时间太长，否则布料会被损坏。

最后的步骤也可能是最无关紧要的——石洗。显然，有些人仍然认为它挺时尚。要取得石洗效果，你需要石头——50 千克石头。这批产品经过了两个小时的循环漂洗。被石头打磨过的牛仔裤又褪掉了一些颜色。

完工后，一个工人将巨大的机器卸载。牛仔裤和石头被共同装上一辆大推车，下一步工序会将他们分开。当牛仔裤从推车中倾倒出来时，很多石头留在车里。一起倒出的石头则会

你知道吗？

牛仔裤所用的面料被认为起源于法国的尼姆，因此得名"尼姆"，后来变成了"牛仔布"。

从烘干机内的洞中筛出。它们被收集起来一次次地重复使用。

最后牛仔裤被熨平。方式是再次对它们充气，但这回使用的是蒸汽。最终完工后。不论被磨损过，漂白过，乃至石洗过，它们将很快风行在你附近的商业街。✏️

.一条牛仔裤有 19 片不同形状的布料
.工人把图样在布料上展开
.按图纸一次剪裁 20 ～ 40 层

4.为保护手指裁工戴着特制手套
5.一条裤子要经过 25 个工人的手
6.女工负责完成裤腿外接缝

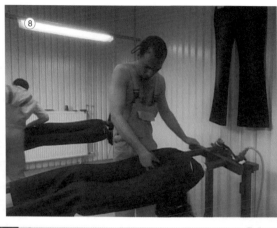

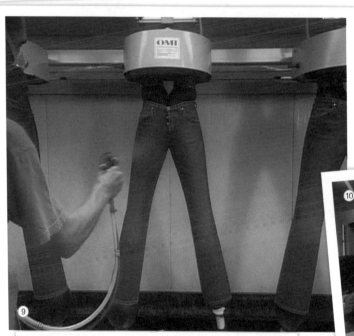

7. 最后手工定好扣子

8. 工人用砂纸磨出自然的褶皱

9. 喷高锰酸钾对裤子漂白做旧

10. 用喷砂打磨做旧

11. 经过 2 小时的石洗

12. 牛仔裤充气烘干并用蒸汽熨平

Jeans

扫描二维码，观看英文视频。

But first where would we be without jeans? Probably all wearing cords. And for those who'd like their denims to be a cut above the rest, here's a factory that makes designer jeans.

There are 19 different pieces of fabric for each pair. First they plot out all the different shapes on a computer to make sure they use every inch of denim. The pattern is printed on a huge sheet of paper for the cutters to snip around. All the pieces are coded to make sure an inside pocket doesn't end up getting stitched onto an inside leg.

Downstairs, they're piling up the denim ready for the template. They cut between 20 and 40 layers at a time depending on how thick it is. This lady takes her work very seriously if the material isn't lying flat the jeans will be botched. Once it's all straightened out they put the pattern on top. Then they can get cutting.

This cutter might look like an extra from Braveheart but he's actually wearing the glove for a very good reason to stop him cutting his fingers off with the compass saw.

The cut denim is taken through to a noisy sewing hall where around 300 people make up to six thousand pairs of jeans a day. They'll use over six miles of denim and a staggering 750 miles of thread, that's nearly enough to stretch all the way from London to Madrid. The outer seams on the legs have to be invisible so they fold both edges over and stitch from the inside. The jeans are handed from station to station with some 25 people working on any individual pair. Finally they add the studs. They have to sew the buttons on the old fashion way. The needlework's complete, and is given a once over from a quality controller.

These jeans are going to be given a distressed look. They're inflated and then workers scuff them with sand paper. It's all done by hand so every pair is a one off. When they get washed later the colour will run from the scuffed areas to make them look worn. Believe it or not this is actually the most expensive part of production. It takes quite a while to get the lines in the right place so they look like natural crease lines.

The next step in making them look like well worn classics is chemical treatment. They bleach them with potassium permanganate which makes the dye fade without damaging the fabric.

More durable denim gets a sand-blasting. The jeans are fixed to clamps and then the unit is sealed so that sand isn't flying all over the place. A worker sprays a pair with about a kilo of sand at 4 bars of pressure. He has to be careful not to spray the same spot for to long or the fabric will be damaged.

Last but perhaps least there's the stonewash look. Apparently some people still think it's still fashionable. To get a stonewash effect you need stones, 50 kilos of them. The batch goes through a two hour rinse cycle and the denim takes a pummeling from the stones removing some of the colour.

When it's finished a worker unloads the giant machine. The jeans and stones are loaded together into a large trolley. They will be separated at the next stage. As the jeans are tipped out of trolley many of the stones are left behind. Those that remain are sieved out through holes in the dryer. All of them are collected and will be reused over and over again. Finally the jeans get pressed. Again this is done by inflating them but this time with steam. And at last they're finished. Whether they've been distressed, bleached or even stonewashed they'll soon be hitting a High Street near you.

Did you know?

The fabric used for jeans is thought to have originated in NImes, France which gave it the name "de Nimes" ("of Nimes") which turned into "denim".

牛仔帽

扫描二维码，观看中文视频。

牛仔是一些勤奋的人。套牛和给牛烙印是繁难的工作。对牛来说也同样辛苦。但牛仔生涯中最煎熬的事情之一，是酷热的沙漠阳光，这就是牛仔帽的由来。怎么样才能做出一顶帽子，约翰·韦恩戴上去也感到自豪？

秘密在于皮毛。优质牛仔帽大部分材质是兔毛，其余是最优质的河狸毛！不错，就是那些牙齿锋利喜欢啃树的啮齿动物，为牛仔帽制作提供了重要的原料。

两种毛必须结合。工序在这里进行，一旦完成后，便被送入这台古老的机器。你可能觉得它看上去就像 20 世纪 30 年代的东西，你是对的。这家得克萨斯州的工厂建于 1938 年，现在生产牛仔帽的过程和当时一样。

毛料落入下面这些巨大的托盘里。你可能认为已经准备就绪了，但且慢，还有个细节问题要厘清。

如果你想得到一顶经得住盗牛贼骚扰的结实帽子，那就需要选用短毛。短纤维互相结合得更好，做出的帽子更结实，因此用这台机器除去较长的毛。长毛从底部排出，而短毛则被留下。

每一顶帽子需要正好 7 盎司混合的兔毛和河狸毛。它们堆积在传送带上，然后用真空泵吸进离心涡轮。当机器旋转时毛会粘在中间的柱子上，并结合到一起。

当所有的毛都粘上了，工人用一块潮湿的黄麻布盖在柱子上，取走帽子浸到 90 摄氏度的水里。如果不这样处理毛就不会粘在一起。水和温度使纤维结合起来。这时你得到一个巨大的圆锥体，更像是哈利·波特的帽子而不是约翰·韦恩的帽子，但目前还是制作的早期阶段，所以请耐心点。

牛仔帽必须结实，所以纤维需要紧密结合。锥体被扭曲成雪茄形，并放置在这些辊子上。揉搓和滚动把纤维编织到一起。每一顶帽子都经过滚压，然后过几次水。在这个过程结束时，材料几乎缩小 60%，但看起来并不很"狂野西部"。在此阶段，它看起来更像是农民戴的帽子，但很快就会变成一顶牛仔帽，只需要再做一点加工。在开始变成正确形状之前，它们需要染色。在 80 摄氏度下经过 2.5 小时，这些帽子被染成深褐色。

所有看过西部片的人都知道，牛仔帽有多种颜色和样式。该给牛仔帽一个合适的形状了。没有边缘的牛仔帽是不完整的，这些机器用于帽子的拉伸和造型。冷水和热水的配合，拉伸和快速干燥，帽子终于开始变成人们熟悉的形状。

除了形状之外，帽子还需要做成不同的尺寸。今天很少有牛仔能够穿上约翰·韦恩的靴子了，更别说他的帽子。每顶帽子都会进入这种机器，通过夹具和热蒸汽的组合把帽子拉大。一旦拉伸完毕，将插入垫铁，用机器挤压到位。

你知道吗？

"十加仑帽子"的大帽檐实际上不能容纳 10 加仑水（1 加仑 =4.546 升）。它们最多可以容纳 6 品脱水（1 品脱 =0.5683 升）。

虽然牛仔在艰苦环境下干活,传统的牛仔帽却有一种天鹅绒般的柔顺质地。刚做好的帽子在这里打磨。生产过程使材料变得粗糙,所以要用砂轮磨出柔软和精美的纹理。唯一的缺点是砂轮在帽子表面上留下一些小毛球,但很容易用火除去。

现在有两个工序,将赋予牛仔帽特有的风格。第一是帽冠形状。一排液压机能同时生产几顶帽子。工人为特定的风格选择一个模具,将这些帽子和模具放入压床后启动。帽子最终现形了,带着世人熟悉3道痕"牧场主风格"。

另外重要的一点是帽檐。帽檐可以相当平坦,或者根据牛仔的个人趣味非常弯曲。有些帽檐可达10厘米宽。

这个公司每年生产50多万顶正宗的牛仔帽,价格从几百美元到上千美元。但你在"狂野西部"工作时,这是为了得到保护所付出的很小代价。

1. 酷热的沙漠阳光是牛仔帽成为必备品的原因
2. 称出7盎司混合的兔毛和河狸毛
3. 短毛被吸进涡轮并粘在柱子上
4. 从柱子上取走帽子放入热水中
5. 通过揉搓滚动使纤维交织在一起
6. 经滚压和过水帽子几乎缩小60%

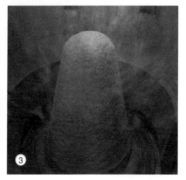

7. 在 80 摄氏度水中给帽子染色

8. 机器完成帽子的拉伸和造型

9. 牛仔帽现出了熟悉的模样

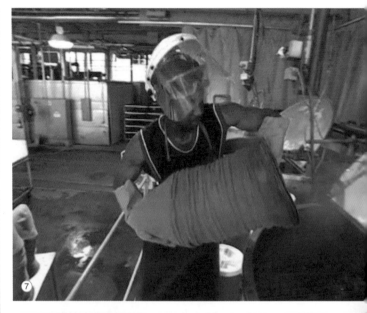

10. 用砂轮磨出精美的纹理

11. 用火烧去磨起来的小毛球

12. 帽冠和帽檐可做成不同样式

Cowboy Hats

扫描二维码，观看英文视频。

The cowboy is a hard working man. Roping cattle and branding them can be really tough work. Not least for the cow. But one of the hardest parts of the job is the scorching desert sunshine and this is where the cowboy's hat comes in. So how do you make a hat that John Wayne would be proud to wear?

The secret is in fur. Much of a good quality hat is rabbit fur, but the rest is finest quality beaver! Yes those toothy little tree-loving rodents provide the important ingredient for making cowboy hats the way it should be.

The two furs have to be combined. That happens here and once it's all mixed up, it's fed into this ancient machine. You may think it looks like something out of the 30's and you'd be right. This factory in Texas was built in 1938 and the process of making cowboys hats is the same now as it was then.

The fur falls into these enormous trays underneath. Now you'd think we would be ready to go, but not just yet. There's one more detail to sort out first.

If you want to get a really tough hat that will survive cattle rustling, you need the short hairs. Shorter fibres bond better and make a stronger hat, so this machine is used to get rid of the longer fur. They are ejected out of the bottom, whilst the shorter hairs are kept in the mix.

Each hat needs exactly 7 ounces of the rabbit and beaver mix. It's piled on to this conveyor and then a vacuum sucks it off into a centrifugal vortex. As it spins round it sticks to the large pillar in the middle and bonds together.

Once all the fur has stuck, the workers cover the pillar in a damp jute cloth and send the hat for a bath at 90 degrees Celsius. Without this treatment the fur wouldn't stick together. The water and the heat help the fibres to bond. Now at this point you've got an enormous cone, more Harry Potter than John Wayne but its still early in the process so hold your horses.

A cowboy hat has to be tough so the fibres need to be well-bonded. The cones are twisted into a cigar shape, and placed on these rollers. The kneading, rolling motion will knit the fibres together. Each one is rolled and then bathed several times. By the end of the process, the material will have shrunk by almost 60% but it doesn't look very "Wild West" just yet. At this stage the hat looks more like something a farmer might wear but this will be a cowboy hat soon, it just needs a bit more work. But before they start being properly shaped they need to be dyed. 2 and a half hours in here at 80 degrees Celsius will give these hats a rich dark brown colour.

Now anyone who has seen a cowboy film will know there are plenty of colours and styles available. Its time to give them their proper shape. No cowboy hat would be complete without a brim, so these machines will help to stretch and shape each one. Combinations of hot and cold water, stretching the material and quick-drying mean the hat finally starts to take on a familiar form.

As well as its shape the hats also need to be made in different sizes. There are very few cowboys today who'd be able to fill John Wayne's boots, let alone his hat. Each one goes into one of these machines which will use a combination of the grips and hot steam to pull it out wide. Once stretched, he will insert the sizing block and the machine will press it into place.

Although they work in a tough environment, the traditional cowboy hat has a soft velvety texture to it. Here freshly made hats are being sanded down. The production process has roughened the material so the sanders are used to produce a soft and almost delicate texture. The only downside is that the sanders leave small felt balls on the hat's surface, but these can be removed easily with fire.

Now there are two stages to giving a cowboy hat its particular style. The first is the shape of the crown. A series of hydraulic presses mean several hats can be produced at a time. The worker selects the mould for a particular style, he then places the hat and mould into the press, and then switch it on. The hat finally emerges, with the familiar three dents of the "cattleman" style.

The other important bit is the brim. They can be quite flat, or extremely curved depending on the cowboy's personal tastes. Some brims can be as wide as 10 centimetres.

Every year this company produces over half a million genuine cowboy hats. Prices range from hundreds of dollars to thousand, but it's a small price to pay for protection when your office is the wild, wild West.

Did you know?

Ten-gallon hats can't actually hold 10 gallons. The most they can hold is 6 pints.

戈尔特斯夹克

说到户外服装，戈尔特斯面料可能是现代最有用的发明之一。1978年它被首次制造，重量轻，能提供良好保护，让穿着者温暖舒适。

那么，这种面料如何能有诸多功能？戈尔特斯服装由两层材料制成。外层通常是尼龙或聚酯，与下面的戈尔特斯膜结合。这层膜每平方厘米有14亿个孔。孔太小，以至雨水无法通过，但却足够让水蒸气或汗液蒸汽跑出去。

第一步是时尚设计，设计师制作一个模板用来裁剪布料。这个软件有助于高效切割材料避免浪费。当设计师绘图构思新夹克时，她的助理正钻进仓库为服装外层选择合适的材料。设计师同时打印模板准备裁剪布料。

把布和模板铺在一起，女裁缝现在可以为夹克所需的不同部分放样。模板被固定好引导刀具切割下面的布料。每块布料裁出以后缝纫就可以开始了。

总共有200多个组件但并非都是材料。还有大量的拉链、纽扣和套锁钉。慢慢将不同部分整合到一起包括戈尔特斯膜。

虽说戈尔特斯膜是夹克真正起作用的部分，外层能够承受面临的各种条件也很重要。第一个测试是磨耗。长期使用和磨损是户外夹克必须经受的。这个装置将布料与粗糙的表面摩擦。摩擦一次相当于大约12小时穿用磨损。每一块布必须承受至少87000次摩擦。少许的损伤意味着这种材料通过了测试，它仍然应该能防水。除了磨损，对材料的强度也进行测试。布料样品被装入一台机器，拉伸到断裂点。以撕开布料所需的力评估材料的质量。

与此同时，夹克的收尾工作也在进行。缝制戈尔特斯材料的一个问题是针脚。和材料的微孔相比针孔可谓巨大，水能够从中流入。然而一个内衬被熨烫到位，能把针脚密封。

戈尔先生于20世纪70年代首次将这种材料开发出来，它的卓越品质能让汗液蒸汽从里面跑出去又防止水从外面渗进来。此外还有重量轻的优点。

但怎么检测它是否防水呢？通过压力。如果有水渗过布料就意味着材料不提供对雨水的防护，穿着者在风暴中会受冷受潮——这显然意味着失败。戈尔特斯织物经历同样的测试会如何呢？虽然施加到两个样品上的力相同，不同之处显而易见。戈尔特斯材料甚至在压力下膨胀起来仍然滴水不漏。夹克差不多完成了，只需要一点额外的工作。纽扣是有用的附件。夹克是防水的但如果不穿好它，也提供不了多少保护。终于质量监控员进行最后的测试，通过向夹克内充满空气可以检查缝合处是否紧密。当工作完成时这件夹克将能经受得住自然考验，即使在天气最恶劣的外部条件下，也能让忠实的旅行者保持衣服内部的舒适干爽。

你知道吗？

马拉松运动员在一次竞赛中，通过出汗和呼吸最多能损失4升液体。

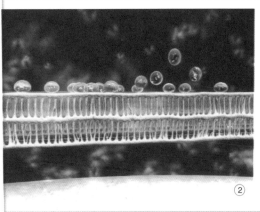

1.戈尔特斯夹克温暖舒适

2.戈尔特斯膜不透水但能让汗液蒸发

3.把布和模板铺在一起

4.一件衣服总共有 200 多个组件

5.将不同部分整合到一起

6.测试中布料有少许损伤但仍能防水

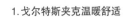

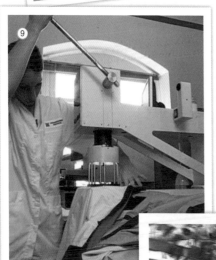

7.通过撕裂实验评估外层强度

8.熨烫内衬解决针脚渗水

9. 压力下检测衣服的防水性

10.用机器钉好纽扣

11.夹克内充气检查缝合是否紧密

12.防雨轻便保暖的夹克完工出厂

Gore-tex Jackets

When it comes to outdoor clothing, Gore-tex is probably one of the most useful inventions today. First manufactured in 1978, it's lightweight, it provides good protection and keeps the wearer warm.

So how does this simple fabric manage to do so much? Gore-tex clothing is made using two layers of material. The outer layer is often nylon or polyester, combined with a Gore-tex membrane beneath. This membrane has over 1.4 billion pores per square centimetre. They're too small for rain to come in, but large enough to let steam or sweat vapours out.

The first step is fashion and a designer creates a template from which to cut the cloth. This software helps cut the material as efficiently as possible to avoid waste. And whilst the designer's drawing up the plans for the new jacket, her assistant is down in the warehouse selecting the right shade of material for the outer layer. The designer meanwhile is printing up her cloth template ready for the cutting process.

With the cloth and the template together, the seamstresses can now lay out the different pieces required to make a jacket. The template is clipped into place and this provides the cutter with a guide for shaping the cloth beneath. With everything cut to size, the tailoring work can now begin.

There are over 200 parts in total but it's not all material. There are plenty of zips, buttons and toggles too. Slowly the different pieces come together including the Gore-tex membrane.

Now although that is the real working part of this jacket, it's important the outer layer can withstand the kind of conditions it will face. The first test seen here is for abrasion. Long term wear and tear is something any outdoor jacket will have to resist. This device rubs the material against a rough surface. Each rub is the same as about 12 hours of wear. Each piece of cloth must endure at least 87,000 rubs. Light damage means this material appears to have passed and should remain waterproof. As well as abrasion, the material's strength is also tested. A cloth sample is loaded into a machine that will stretch it to breaking point. The force required to tear it is then measured to assess the material's quality.

Meanwhile the finishing touches are being added to the jacket. One problem with sewing in the Gore-tex is the stitching.

Water can get in through these holes which are huge compared to the material's microscopic pores. However, a liner is heated into place which seals the stitching.

First developed in the late 70's by Mr Gore, the material's unique properties that allow sweat vapour to escape from the inside whilst preventing moisture coming in from the outside has the added bonus of being lightweight.

But how can they test if it is waterproof? By using pressure. If water seeps through the fabric it means the material offers no protection against rain and the wearer would end up cold and wet in a storm... this is clearly a fail... But when the Gortex fabric is exposed to the same test? Although the force applied to both samples is the same, the difference is clear. Even when bulging under the pressure, the Gore-tex remains watertight.

The jacket is now almost complete, it only needs a few extras. Buttons are a helpful addition. The jacket may be waterproof, but if its not done up, it won't offer much protection. Finally quality control examiners run one last test. By pumping the jacket full of air, they can check that the seams are secure. When the work's done this jacket will resist the elements, to keep the dedicated rambler dry on the inside, even when the weather is at its worst on the outside.

Did you know?

Marathon runners can lose up to 4 litres of fluid in a single event through sweating and breathing.

羽绒服

扫描二维码，观看中文视频。

冬天很少有什么比一件鹅绒外套更让你温暖了。要做出鹅绒外套，这家工厂只用最细小最蓬松的内层绒毛做原料。这些天然保暖材料非常柔软，被称为羽绒。

羽绒在机器筛子里旋转飞舞。尘土和破碎的羽毛掉到地上。只有最完好的羽毛被吸入管子里送进旋转洗涤机。一种非常温和的洗涤剂在40摄氏度的温度下洗清羽绒的尘污。几乎20千克的污垢和油脂从羽绒里洗出来。

现在该把它放入烘干机以免羽绒腐烂变质。羽毛在烘干机里被100摄氏度的热风吹干。20分钟后干透了就可以使用。但在此之前还要按大小进行分拣。这个机器里有一股气流把羽毛吹起来。气流的强度刚好能够抬升起较轻的羽绒，较大的羽毛就会掉下来。同样的事情发生在第二个箱子里，这次是为了收集中等大小的羽毛。中等大小的羽毛被送去填充枕头，更大一些的羽毛注定会被用作肥料。

15分钟后羽毛被分拣完毕。最小最珍贵的羽绒在机器的顶端采集，它们被放到袋子里送去做外套。一位针线工做出口袋，装进羽绒填充。每个口袋里填装羽绒的量是非常精确和预先设定的。这使得外套看起来匀称并足够保暖。这个袋子的填装量是8克。每件羽绒外套有18个这样的袋子。

一块大片的衣料上预先标记出做外套相应的部分，被仔细地裁剪下来，用作衣服的外层。这是一种特殊的多孔材料，既透气又防水。因为羽绒服的外层由100多个不同的部分构成，仅把它们从衣料上完全剪下来，就需要花费40分钟。

另一个针线工用模板在布料上画好图样，然后按线条对外套进行缝纫。她们用尼龙线把装着羽绒的口袋缝到外套上。然后她们用交叉缝法，沿着早先画好的图样缝制。这种方法可防止羽绒在口袋里流动和缩团，还可以让外套整体看起来蓬松而匀称。当所有部分都制作成功后，就该把它们缝合到一起了。在这项工序中，衬里、口袋和拉链都会被加在外套上。

经过一个半小时的精心缝制，外套完成了。因为做一件外套需要花不少时间，这家工厂一天仅能生产200件。但他们仍会用掉1300万根羽毛。

你知道吗？

第一次世界大战中，妇女给所有未穿制服的男人分发白色羽毛，以嘲讽他们的怯懦。

1. 冬天里一件鹅绒外套让你温暖
2. 用最细小蓬松的内层绒毛做原料
3. 清洗羽绒上的污垢和油脂
4. 气流中最小的羽绒飘到上面
5. 女工准备缝制装羽绒的口袋
6. 称出每袋所需羽绒的重量

7. 羽绒定量灌装到口袋里
8. 外层面料为特殊的多孔材料
9. 把羽绒口袋缝到相应的面料上
10. 交叉缝法防止羽绒流动和缩团
11. 衣服的前后片缝合到一起
12. 最后装好衬里、口袋和拉链

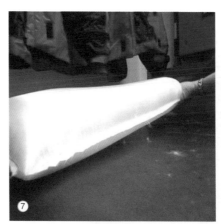

⑦

⑧

⑨

⑩

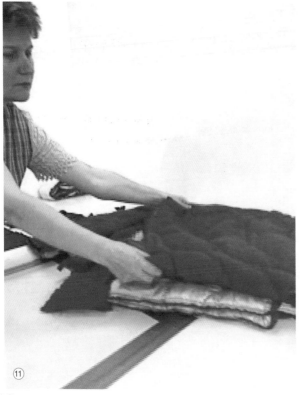

⑪

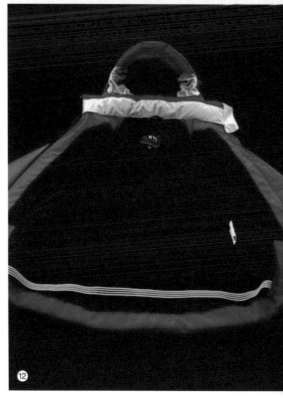

⑫

Down Jackets

There are few things that will keep you warmer in the winter than a coat made with goose down. To make them, this factory only uses the smallest and fluffiest inner feathers. These natural insulators are super soft and are known as down.

The down is whirled around in a sifter. Dust and any broken feathers fall to the floor and only the healthies feathers get sucked up the pipe on their way to a spin wash. A very mild detergent is used at forty degrees to get all the dirt out. Nearly twenty kilos of dirt and grease has come out of the feathers.

And now it's on to the dryer so the feathers won't rot. The feathers are blown around these containers at 100 degrees Celsius. After twenty minutes they're dry and ready to be used. But first they've got to be sorted by size. There's an air stream in the machine which blows the feathers around. The stream's only strong enough to lift the lighter down so the larger feathers fall to the ground. The same thing happens in the next box - this time to catch the medium feathers. The medium sized feathers go off to be stuffed into pillows and the larger ones are destined to be used as fertiliser.

After 15 minutes the feathers are sorted. The smallest most valuable down has collected at the top, it's put into a sack and taken off to make the jackets A seamstress makes the pouches that will contain the down stuffing. Each pouch is filled with a precise and pre-determined amount of down to make sure there's an even look to the jacket and enough insulation. For this pouch that's eight grams. Eighteen of these bags go into every one of the down jackets.

A single, large piece of fabric is marked out to make up the outside of the coat before being carefully cut. It's a special porous material that lets the jacket breathe while still being waterproof. With over a hundred different parts needed to make up the outside of the coat it takes forty minutes just to cut them all out.

Another seamstress stencils a design on to the fabric, this'll be stitched around to make the coat quilted. They use a nylon thread to attach the pouches to the coat. And then they sew in cross seams along the design that was sketched out earlier. These stop feathers from collecting in one part of the pouch and keep the jacket looking puffy all over. All the pieces are ready and it's time for them to get sewn together. It's here that the lining, pockets and zips are attached to the jacket.

An hour and a half of careful sewing later and the jacket's complete. Because of all the time involved in making one this factory only produces two hundred a day, But they still manage to get through 13 million feathers.

扫描二维码，观看英文视频。

Did you know?

In world war I women gave out white feathers as a symbol of cowardice to any man who did not wear a uniform.

开司米羊绒

扫描二维码，观看中文视频。

开司米羊绒被认为是做衣服的最柔软、最舒适的羊毛。但价格也最为昂贵，其中自有原因。

开司米羊绒并非如你想象的来自绵羊。它出于山羊，是的，山羊，那些绵羊的堂兄弟。它们散发着臭味，长着粗毛，在偏远山坡上游荡。

开司米羊绒真正精华的秘密，是山羊的内层绒毛。山羊只在世界上特定的地方生长这种柔软的细毛，那里气候的寒冷程度，足以让山羊长出这种毛来。

像鸭绒一样，山羊生长出这种绒毛用来隔热保温，它的特别之处在于结构。山羊的外层毛很硬实。在显微镜下看起来相当粗糙和不平。

内层绒毛完全相反，它要平顺光滑得多。这就是为什么开司米羊绒质地柔软，与皮肤接触时感觉也舒适得多。

收集羊毛不容易，山羊并不喜欢自己的毛被拔掉。一旦采集好羊绒，便送到当地工厂纺织成布。这个是穿着开司米羊绒的最好办法。专家们知道该怎么处理刚收取的羊毛。首先需要将它们混合。每个山羊的毛都有不同色泽，经过混合后，工厂整个批次的羊毛便有了均匀的颜色。

然后，它们需要从这台老式机器里走一遭。这个过程被称为"梳理"。羊毛经过梳子后朝同一方向排列，单股的绒毛结合起来像一大张布。从这排辊子远端过来的羊毛布，被送到下一流程纺成线。

这种样式的线很柔软，手感极佳，但强度非常弱。需要再次细纺。最后工序增加了纤维强度，1千克的羊绒可以生产出超过6000米的开司米羊绒线。

英国只有50个左右的开司米羊绒生产厂家和总共大约2500只山羊。在英国，大多数用于服装生产的开司米羊绒是进口的。

就像工业革命时代古老的纺织作坊，开司米羊绒在织布机上生产出来。熟练工人把新的羊毛线装在梭子上，准备开始编织，虽然是一台旧机器，速度之快仍让人眼花缭乱。

这种织布机的设计非常古老，但织起布来却相当有效。一个预先装有羊毛线的高速梭子在两排线之间飞快地来回穿行。梭子上的羊毛线就像脚手架结构，把周围其他羊毛线紧固在一起。

织一条围巾需要3只山羊的羊绒，并且在机器上花费4小时。如果要做一件毛衣，大约需要6只山羊的羊绒。

对于传统的开司米羊绒围巾，还剩下唯一的事是两端蓬松松的流苏。这是机器做不了的，需要大量人工来打绳结，并赋予这些围巾著名的外观。

下次你把柔软的开司米羊绒围巾戴在脖子上的时候，请想想老公山羊温柔的一面吧。

你知道吗？

中国科学家利用大豆纤维制造了一种人造羊绒。这种材料像丝绸一样柔软，成本仅为羊绒生产的1/15。

1. 柔软舒适而昂贵的开司米羊绒围巾
2. 开司米羊绒出自山羊的内层绒毛
3. 只有特定地方的山羊生长这种细毛
4. 显微镜下对比，内层绒毛平顺光滑
5. 对每个批次的羊绒进行混合
6. 羊绒经过梳理朝同一方向排列

7. 单股的绒毛结合成一大张布

8. 每千克羊绒生产 6000 米绒线

9. 把线装在梭子上进行编织

10. 织布机工作示意

11. 3 只山羊的绒可以做一条围巾

12. 手工做出两端蓬松的流苏

Cashmere

Cashmere is considered to be one of the softest and most comfortable wools that can be used for clothing. But its also one of the most expensive, and here's why.

Cashmere doesn't come from sheep like you might expect. It comes from goats, Yes, goats. Those smelly, coarse-haired cousins of sheep that like hang around remote hillsides.

The secret to a really good crop of cashmere is the goat's undercoat. They only produce this downy under-hair in certain parts of the world, where the climatic conditions are cold enough to encourage it's growth.

Like eiderdown from ducks, this under-hair is produced to insulate the goat, but what makes it so special is its structure. The overcoat is made up of thick coarse hair. Seen up close under a microscope, it would appear quite jagged and rough. This is the complete opposite of the undercoat which is far smoother. It's what gives cashmere its soft texture and makes it so much more comfortable against the skin.

It isn't easy collecting the wool because the goats don't really like having their hair pulled. Once a crop has been collected, it's sent off to the local mill to be turned into cloth. No, this isn't the latest way to wear cashmere. These workers are experts and know what to do with the fresh wool. First it needs to be blended. Every goat's wool is a different shade so by mixing them together the factory gets an even colour throughout the batch.

It then takes a bit of a journey through this old machine. This process is known as "carding". Combs brush the wool so that it all runs in a similar direction and the separate strands bind into a big felt-like cloth. The sheet of wool that emerges from the far end of this row of rollers is taken down to the next stage of the process to be spun into thread.

In this form the thread is very soft and wonderful to the touch, but it's also very weak. It needs to be spun again. This last stage strengthens the fibre and 1 kilo of wool can now produce over 6,000 meters of cashmere thread.

The UK can only boast around 50 cashmere producers with a total of about 2500 goats in all. Most of the cashmere used in the UK for clothing production is imported.

Like the old cloth mills of the Industrial Revolution, the cashmere is woven on a loom. Experienced hands load up the shuttles with fresh wool, ready to start the knitting process and although it is an old machine its faster than the human eye can follow.

This kind of loom is a very old design but very effective for weaving cloth. A high speed shuttle, which was pre-loaded with wool, is passed very quickly back and forth between two rows of threads. The thread from the shuttle works like a scaffolding structure to bind the other threads around.

It takes the wool from 3 goats, and 4 hours on one of these machines to weave a scarf. But if you want to make a sweater you'll need the wool from about 6 goats.

The only thing that remains for the traditional cashmere scarf is the loose tassels at the ends. This is something the machines can't do but there are plenty of spare hands to help tie knots and give the scarves their famous look.

So the next time you wrap that soft cashmere scarf round your neck, spare a thought for the softer side of the old Billy Goat Gruff.

Did you know?

Chinese scientists have created an artificial cashmere using soy bean fibre. The material is as soft as silk and cost just 1/15th of what cashmere does to produce.

袜子

2005年，在英国卖出超过6亿双的袜子。畅销的袜子除了舒适以外，还需要时髦。所以在袜子织出第一针前，设计师需要拿出配色方案和构图样式。

绚丽的袜子并不是现代社会的发明。1000多年前，野蛮人就穿着鲜艳的袜子意气风发地投入战斗。一旦设计师对图案式样满意后，就可以选择具体的颜色。这需要一些时间，仅在这个工作室里，就有超过500种不同的色调可供挑选。

选色完成后就可以在电脑上做模拟设计。电脑会为她进行工作，计算出环形的宽度和数量，颜色的顺序，甚至线圈的多少。线从这个仓库采集而来。它们有170个不同类型，由羊毛、棉、羊绒、丝、混合线和化学纤维制成。他们每年要使用1500吨线，所以细微的款式变化都是设计的成功。

在缝纫间里400台编织机都在不停地制作袜子。一天能生产45000双之多！

一个工人把线头搭上机器，然后插入数据卡——其中包含了设计方案。编织机1分钟之内能绕27000圈，即使是规模最大的妇女缝纫协会也难以匹敌。肉眼看来只是模糊不清的一团色彩，但通过慢镜头你能大致看到160根针如何把来料织成袜筒。5分钟后袜筒达到足够长时，确切说来有66厘米，就被吸进一个圆筒去往下一站。

袜子是里朝外缝制的，需要翻转过来。一个真空装置能完成这项任务，遗憾的是，放置袜子的这项重复工作还只能是手工活。袜子快要做成了，但还不能送到街上去卖，袜子的脚趾部位仍是一个大洞。每只袜筒都要仔细安放到一排针上，这个工作需要全神贯注、手法稳定及圣徒般的耐心。

一台机器把多余的织物剪掉，袜筒被缝到一起。待到脚趾部分合上后，袜子在130摄氏度下进行蒸汽熨烫，7秒钟便足够了。时间长一点就可能烧坏袜子。

没有皱褶，袜子看去崭新并且穿着舒适。这是早期的袜子无法比拟的。大约3000年前希腊人的袜子用乱糟糟的动物毛发做成！袜子出厂之前质量监控人员会进行再次检查。最后两只袜子被缝到一起，确保它们成对成双，至少当下如此！✏️

你知道吗？

双脚上有52块骨头，约占人体全部骨头总数的1/4。

1. 织袜前首先拿出方案构图
2. 电脑计算出针数及其他数据
3. 根据设计从仓库选取线材
4. 线头搭上机器后插入数据卡
5. 编织机 1 分钟能绕 27000 圈
6. 慢镜头下的 160 根针在工作

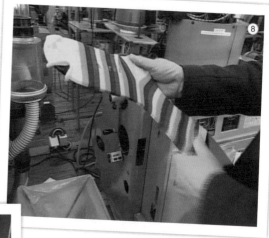

7.真空装置把朝外编织的袜子翻转

8.织好的袜筒脚趾部分是一个洞

9.脚趾开洞位置安放到一排针上

10.缝合好脚趾部分

11.袜子将进行蒸汽熨烫

12.最后一针确保袜子成双成对

Socks

扫描二维码，观看英文视频。

In 2005 over six hundred million pairs of socks were sold in the UK. As well as being comfortable, bestsellers also have to be fashionable. So before a sock's first stitch a designer works out a colour scheme and pattern.

Flashy socks aren't a modern day invention. Barbarians were steaming in to battle wearing brightly coloured socks well over a thousand years ago! Once the designer's happy with the pattern she can pick out the exact colours. This can take some time, there are more than 500 different shades to choose from in this studio alone.

That done she can model her design on a computer. It does the maths for her, calculating the width and number of rings, sequence of colours and even the number of loops.

The threads are collected from this warehouse. They've got 170 different types, made from wool, cotton, cashmere, silk, mixed threads and chemical fibers. They use 1,500 tons of it every year so a bit of variety does go a long way.

Over in the sewing hall 400 knitting machines are constantly churning out socks. They knit a whopping forty five thousand pairs a day!

One of the workers mounts the threads onto a machine, and then inserts a data card which contains the design. The knitting machine gets through twenty seven thousand loops in just a minute! Even the largest of sewing circles would struggle to match that pace. It's just a blur of colour to the naked eye, but in slow motion you can see some of the 160 needles that are knitting the fabric into tubes. After about five minutes they're long enough, 66 cm to be precise. They're sucked through a cylinder towards the next station.

They've been stitched inside out and need to be re-versed. A vacuum helps but unfortunately the repetitive task of mounting them has to be done by hand. The socks are almost finished but they're not ready to hit the High Street yet, there are large holes where the toes should be. Each loop of each sock has to be carefully mounted onto a row of needles. It's a job that requires concentration, a steady hand and the patience of a saint.

A machine cuts off the spare fabric and sews the loops together. Once the toes are closed, the socks are steam-pressed at 130 degrees Celsius. 7 seconds is all it takes, any longer and they could get singed.

Crease-free they'll now be fresh and comfortable to wear. That's more than can be said for the earliest known socks. The Greeks made them nearly 3,000 years ago out of matted animal hair! Before they head out of factory door a quality controller gives them a once over. Finally they are paired up and sewn together to make sure that they stay in pairs… for now!

Did you know?

The 52 bones in the feet make up nearly a quarter of all the bones in the human body.

威灵顿雨靴

扫描二维码，观看中文视频。

看来又要下雨了。英国经常下雨，这也许能解释为什么英国人发明了世界上最著名的防水靴之一——威灵顿雨靴。全橡胶制成的现代威灵顿雨靴适用于不同环境，从化工厂的紧急状况，到一般的泥泞场地。

最早的威灵顿雨靴是威灵顿公爵发明的，由皮革制成。后来人们用温暖热带气候中生长的橡胶树液做原料。

然而，生产天然橡胶非常昂贵，因此制造商转而使用价格低廉的人造合成原料，比如聚氨酯来替代。大桶里的化学物品充分混合在一起。当它们结合后充分静置，就成为制作结实靴子的理想材料。

为了形成不同的色彩，在制造合成橡胶的化学过程开始前，便要往混合物中添加颜料。在装有车轮的大桶里，工人把一批大约1000升的原料混合起来。

再把这种混合化学物品非常小心地运到工厂车间。在这里它们被送到靴子生产机器上，工序就可以开始了。那么靴子是如何制作出来的呢？

实际上这是个相当简单的过程。液体聚氨酯倒入一个靴子形状的模具里，化学反应会导致橡胶样的材料膨胀。然后凝固成靴子的形状。但如果你这样做靴子，做出来的靴子是实心的。你完全无法把脚放进去，于是这些模型腿脚被引进了生产过程。

金属腿脚有各种尺寸。插入模具后，聚氨酯材料将围绕假脚形成靴子。这便是生产过程的核心。首先工人把溶剂喷到模具里，使模具像不粘锅一样，防止聚氨酯混合物粘在上面。接着操作员把长袜放置在模具中。形成靴子的内层，使它穿起来更舒适。有些型号需要加入钢制的鞋头。最后加入内鞋底，能使靴子更跟脚，接下来机器开始工作。模具密封后，接上一根管子，开始泵入聚氨酯化学混合物。

仪表用来监控压力，检查员也需要定期采样，以确保混合物正常。仅此一家工厂每年可以生产60万双靴子。当机器打开后，带有现成鞋底的新靴子就出现了。因为产量大，机器人被用来提高生产率，但大部分工作仍然靠手工完成。

新做好的威灵顿雨靴从模具中取出时有粗糙的边缘，所以要进行修剪，任何多余的橡胶边角被除掉。然后将靴子送去进行质量检查和测试。

多功能防水靴不仅用来踩泥坑。它们能在各种极端环境下工作，因此需要经受磨损和撕扯。测试包括弯曲鞋底10万次，以确保它能经受这么长的行走距离。

接着对钢鞋头进行测试。它们能在重物落下时保护脚趾，必须能经受20千克重物的冲击。靴子本身必须耐久而富有弹性。材料样品

你知道吗？

橡胶雨靴最早是在苏格兰批量生产的。工厂老板相信恶劣的天气能促进销售。

被拉伸到 4 倍以上长度，来测定他们能承受多大的力。

如果一个批次通过了所有必要的质量检测，就印上公司的商标，准备发送到商店。就像传奇的威灵顿公爵在滑铁卢打败拿破仑一样，以他的名字命名的靴子，也无疑能在保护双脚干燥的战斗中取得胜利。

1. 威灵顿雨靴现用合成原料制成
2. 化学原料在大桶里混合静置
3. 不同尺寸的腿脚模型
4. 威灵顿雨靴生产车间
5. 将溶剂喷到模具中
6. 长袜套住腿脚模具形成靴子内层

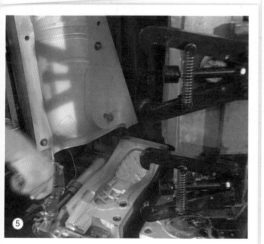

7.放入腿脚模具和内鞋底

8.压进聚氨酯混合物并检查凝固时间

9.修整雨靴粗糙的边缘

10.雨靴进入质检

11.测试雨靴经受磨损和撕扯

12.通过质检后印上公司的商标

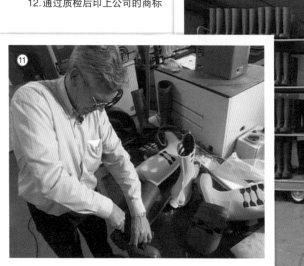

Wellington Boots

Looks like rain again. We get a lot of that in the UK which may explain why it was an Englishman who invented one of the world's most famous waterproof boots, the Wellington. The modern welly is an all rubber affair that is used in a variety of environments, from chemical factories and emergency situations to the bog-standard muddy field.

The original Wellington boot was invented by the Duke of Wellington but was made out of leather. Later they were made using the resin of the rubber tree which grows in warm tropical climates. However, producing natural rubber became very expensive so manufacturers turned to cheaper synthetic alternatives like polyurethane. The chemicals contained in these enormous vats will be mixed together vigorously. When they combine and set they will create the ideal compound for a sturdy pair of boots.

For a variety of different shades, colouring is added into the mixture before the chemical process that makes the synthetic rubber begins. Using large drums on wheels, the staff will mix a new batch of about 1000 litres of the raw material.

This chemical concoction will then be very carefully transported to the factory floor. Here, it's hooked up to the boot production machine and the process can begin. But how are these boots made?

Well, it's actually quite a simple process. The liquid polyurethane will be poured into a boot-shaped mould where a chemical reaction will cause the rubbery material to expand. It then solidifies into the boot shape. However, if you made boots like this, they'd be solid. You wouldn't actually be able to get your feet into them. But that's where all these spare legs come into the process.

These metallic feet come in all sizes. Once inserted into the mould, the polyurethane material will form the boot around this substitute foot. And this is the heart of the production process. First the workers spray solvent into the mould. This works like a non-stick frying pan and stops the polyurethane mixture from sticking. Next, the operator will place a stocking in the mould. This will form the inside of the boot and make it more comfortable to wear. Some models call for steel tip toe caps which are fitted next. And finally an inner sole is added to give your foot grip inside the boot and the machine is put to work. Once the mould is sealed, a pipe is attached and the polyurethane chemical mix is pumped in.

Gauges monitor the pressure and an inspector also takes regular samples to ensure the mixture is right. This single factory can produce about 600,000 pairs of boots every year. When the machine opens up, brand new boots emerge complete with a handy sole. As turn-over is high, robots are used to improve productivity but much of the work is still done by hand.

The new Wellingtons will have rough edges when they emerge from the mould, so the boots are trimmed, and any extra rubber nubbins removed. They're then passed on for quality control to inspect and test.

These versatile water proof boots aren't just used for stomping through puddles. They will end up working in a variety of extreme environments, so they need to withstand abrasive substances as well as wear and tear. Testing includes flexing the sole through over 100,000 paces to be sure it can last the distance. Next the steel tips are tested. They should protect the toes from heavy objects being dropped on them. They have to survive the impact of a 20 kilogram weight. And the boots themselves must be durable but flexible. Material samples are stretched to over 4 times their normal length to determine how much stress they can withstand.

When the batch has passed all the necessary quality controls, they're marked up with the company logo ready to be sent out to the stores. So just like the legendary Duke who defeated Napoleon at Waterloo, the boots that bear his name promise victory in the battle to keep our feet dry.

Did you know?

Rubber wellies were first mass-produced in Scotland. The factory owner believed that the poor weather would boost sales.

口香糖

电动牙刷

牙线

牙膏

牙齿修复

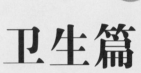

卫生篇

隐形眼镜

创可贴

针头

橡胶手套

天然发刷

一次性尿布

口香糖

扫描二维码，观看中文视频。

它们味道很好，有助于消除口臭。有人甚至用口香糖代替牙刷，你知道是为什么吗？

糖会增加热量摄入并引起龋齿，因而现在许多口香糖使用木糖醇之类的替代品。它们来自天然原料比如甜玉米，可以改善味道，还有无糖的其他好处。稍后再细说。

口香糖的生产，始于薄荷脑晶体和薄荷油。这是活性很强的混合物，有极高的挥发性，工人在搅拌时需要戴上防护眼镜。接下来，加进无味橡胶和蜡混合物做成的小球。这就是口香糖的弹性基料。

但口香糖是如何保护牙齿不长菌斑的？当你吃东西后食物残留在牙齿上，给细菌层提供营养。这层细菌被称为牙菌斑。细菌产生乳酸这是损害牙齿和产生龋齿的主要原因之一。胶母会除掉一些食物残留，但真正起作用的是无糖甜味剂。

科学家在口香糖里加进木糖醇后，实现了这一突破。他们发现酸值明显降低，龋齿也相应减少。木糖醇和一些人造甜味剂加进混合物中。单靠木糖醇不如真正的砂糖甜，所以需要额外的措施。最后把薄荷油、薄荷醇等调味剂加入混合物中。

最终做出来是真正的大块硬糖，一个 100 千克重的口香糖球。这一大块足以变成约 55000 块单独的口香糖。在口香糖变硬之前要抓紧切成小块。巨大的口香糖块被辊压成一大张，然后通过专用切刀，变成熟悉的口香糖形状。下一步将它们分开。此时口香糖已经能吃了，但还缺少一个重要的点睛之笔。

口香糖被送进一台像洗衣机那样的混合器。在这里口香糖被覆盖上多层材料。首先是液体接着是粉末，然后是更多的液体。总共有 40 个超薄涂层组成光滑甜脆的包衣。大约 5 小时后，混合完毕。做出来的小块现在看上去更诱人，更像你在一包口香糖里看到的那样。成品口香糖现在可以咀嚼了，并能帮你防止蛀牙。

通过实验室测试能证明口香糖的保护功能。测试者嘴里装上测量 pH 值或酸度的设备。首先她喝下细菌培养液。图表显示 pH 值的卜降。这意味着酸度提高易于引起龋齿。现在给她口香糖看看对酸度有何影响。第一个品种含有木糖醇。马上有图表指示 pH 值显著增加，这意味着更低的酸度。木糖醇作为中和剂发挥了作用。接下来是普通的无木糖醇口香糖。它的效果由图中红色曲线显示。开始时酸的水平相同，但它不像木糖醇口香糖对酸的中和那么有效。

然而尽管这种甜味剂口香糖对牙齿有好处，如果大量摄入的话会有轻泻的作用。牙医建议保持牙齿健康的最佳途径是经常刷牙和使用牙线。但这种口香糖显然也是另外一种有效武器。回到口香糖工厂，成品口香糖进行分拣然后打包。从味同嚼蜡的橡胶球，到保护牙齿的额外防线，口香糖让你的微笑和呼吸变得甜蜜。

你知道吗？

一位布兰妮·斯皮尔斯的铁杆粉丝，花了 7000 英镑购买了一块二手口香糖，该口香糖曾被一位潦倒的流行歌手嚼过。

1. 口香糖中的木糖醇来自甜玉米
2. 搅拌薄荷脑晶体和薄荷油
3. 无味橡胶和蜡做成口香糖基料
4. 加进木糖醇和人造甜味剂
5. 加进薄荷醇等调味剂
6. 做出 100 千克重的硬糖

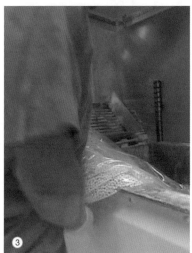

7. 把大块分割成小块并压成饼状
8. 平均分切成小方块
9. 在混合器中喷上多层材料
10. 实验者测试口香糖对口腔 pH 值的影响
11. 监测屏幕显示，木糖醇口香糖可降低口腔酸度
12. 口香糖进入包装车间

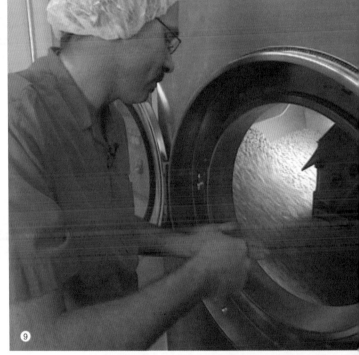

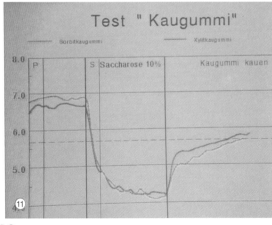

Test " Kaugummi"

Sorbitkaugummi　　　　Xylitkaugummi

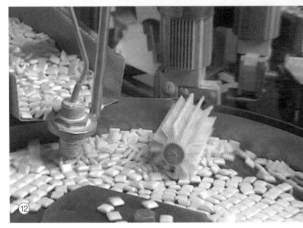

Chewing Gum

扫描二维码，观看英文视频。

They're tasty and they're just the thing to help get rid of bad breath. Some people even use gum as an alternative to the tooth-brush, and you're about to see why.

Using sugar would add calories and causes decay. Today many gums are made using sugar substitutes like Xylitol. Made from natural sources such as sweet corn they add taste and other sugar-free bonuses. More about that later.

Gum production starts with menthol crystals and peppermint oil. It's a potent combination and the vapours are so strong, workers have to wear protective glasses whilst stirring it. Next, beads made out of a tasteless rubber and wax combination are added. This forms the chewing gum's elastic base.

But how does chewing gum protect your teeth from plaque? Well, when you eat, food deposits left on your teeth feed a layer of bacteria in your mouth. This layer is known as plaque. The bacteria produce lactic acid and this is one of the major causes of damage and decay. Gum may remove some food deposits, but the real benefit comes from the sugar-free sweetener.

Scientists made the breakthough after adding Xylitol to chewing gum. They found a noticeable decrease in the acid levels and therefore less decay.

The Xylitol is added into the mix as well as some artificial sweetener. Xylitol alone isn't as sweet as real sugar, so an extra kick is needed. Finally the peppermint menthol flavouring is added in to the mix as well.

What emerges is a true gobstopper, a 100 kilogram ball of chewing gum. This enormous lump is enough to produce about 55,000 individual pieces. Before the gum hardens it must be swiftly cut down to size. The enormous gumball is rolled out into a large sheet and then passed through a special cutter which produces those familiar chewing-gum shapes. Next they're all separated. At this stage the gum is edible, but missing an important finishing touch.

The pieces are fed into this washing machine like mixer. Then multiple layers are added. First liquid, then powder, then more liquid. 40 micro-thin layers in all are built up to produce that smooth, sweet crunchy coating. After about 5 hours, the mixing is complete. The resulting pieces now look much more appetising and more like what you would get in a pack of gum. The finished gum is ready to chew, and help prevent tooth decay.

Its protective qualities are demonstrated in a laboratory test. The subject's mouth has been fitted with a device which measures pH or acidity. First she drinks a bacteria-feeding solution. The graph shows a fall in the pH value. This means higher acid levels which contributes to tooth decay. Now she's given chewing gum to see what affect it has on acidity... The first variety contains the Xylitol. Immediately, the graph registers a marked increase in pH value. This means lower acidity. The Xylitol is working as a neutraliser. Next the ordinary Xylitol-free gum. Its effects are seen in red on the graph. To begin with the acid levels are the same, but it doesn't neutralise the mouth as much as the Xylitol gum.

However, while this sugar-substitute gum maybe good for the teeth, it can also have a laxative effect if taken in large amounts. Dentists advise that the best way to keep teeth healthy is with regular brushing and flossing. But this gum is clearly another useful weapon.

Back in the gum-factory, the finished pieces are sorted out so they can be packaged up. From a ball of tasteless rubber to an extra line of oral defence that helps keep your smile and breath sweet.

Did you know?

A die–hard Britey Spears fan paid around £7,000 for a used piece of gum that had been chewed by the troubled pop star.

扫描二维码，观看中文视频。

电动牙刷

"一天一苹果，医生远离我"。而让牙医远离的最有效办法，就是一把先进的电动牙刷。

电池盒子的诞生，始于几百万个微小的塑料颗粒。这些塑料颗粒被熔化成黏稠的液体，然后注入模具，在那里硬化成牙刷的外壳。一台计算机对它进行扫描，以确保没有丝毫缺陷。任何微小的瑕疵，都可能导致进水，破坏即将装入其中的电动设备。

这些是牙刷头，此刻还是光秃的，但马上会得到延伸。刷毛由一种叫作聚酰胺的人造纤维制成。不同颜色显示刷毛不同的粗细。粗的刷毛被安放在刷头的中间，而在刷头边缘，会使用更细更软的刷毛，以避免损伤牙龈。

只需1秒钟的时间，刷头就能装上刷毛。通过慢动作，你能看到是如何操作的。他们用银色的金属丝将刷毛固定到位。刷毛折叠在一小段金属丝上，此后被塞进刷头的扎洞里。刷头在旋转，所以机器能一动不动在整个刷头上装好刷毛。

现在牙刷已经有了丰满的刷头，但刷毛的边缘还有些粗糙，所以需要"理发"。一个锋利的刀片切掉任何散乱的刷毛，使得每根都达到正确的长度。最后，他们钝化刷毛的锋利边缘。

工厂的另一边正在安装牙刷柄。当金属针到位时，光传感器会告知机器，迅速插上一个塑料套柄，用于安放牙刷头。接着进行紫外线照射，将细菌杀灭，使塑料完全消毒。用电脑进行扫描，再次确定达到无菌标准。

下一步就要装上牙刷头了。到目前为止，壳体还是空的。但很快，马达和变速器就会被装进电动牙刷。左边是变速器，右边是马达。机器将它们钳到一起并且固定。接着和一个可充电电池焊到一起。工人确定电路连接良好后，电器元件就会安装到壳体里。只需简单一扭，塑料塞就将电池密封在壳内。

刷头以每分钟4万转的极快速度旋转。并且还会在适当范围轻轻地来回摇晃。

一名质检人员给牙刷做最后一次快速检查。当刷头和刷柄安装到刷把后，其中的一些被拿去进行测试。它们会在接下来的320个小时内持续刷一副假牙，这远远超过5年的刷牙时间。

如果样品测试通过，这批牙刷将被认定达标，并附上5年的质量担保出售。只要没有掉到马桶里，它们可能会刷上很长的时间。

你知道吗？

4世纪，埃及人发明了一种最早的牙膏。它是由盐、胡椒、薄荷和干鸢尾花瓣制成的。

- 远离牙医的最好办法是使用电动牙刷
- 对外壳扫描检测确保无瑕疵
- 不同颜色显示刷毛不同粗细
- 刷毛折叠在小段金属丝上塞进孔洞
- 对刷毛的锋利边缘进行钝化
- 金属针到位时插上塑料套柄

7. 紫外线杀菌后装牙刷头
8. 装上牙刷头
9. 马达和变速器准备完毕
10. 确定电路连接良好
11. 将电器元件安装到壳体里
12. 连续 320 个小时刷牙测试

Electric Toothbrush

扫描二维码，观看英文视频。

An apple a day keeps the doctor away, and one of the most efficient ways to keep the dentist at bay is a state of the art electric toothbrush.

The battery cases start out life as millions of tiny plastic granules. These are melted down into a sticky liquid and then injected into a mould where they harden into the shape of the casing. A computer scans it to make sure there aren't any flaws. If there is even a tiny imperfection water could get in and damage the electrics which will go inside later.

These are the brush heads. They're bald at the moment but they're about to get some extensions. The bristles are made from a man made fibre called polyamide. The different colours indicate the thickness of the bristle. Thick ones go in the middle of the head, and on the edge they use thinner softer ones so as not to damage gums.

A head is stuffed with bristles in just a second. In slow motion, you can see what's going on. They use a silvery wire to hold the bristles in place. The bristles are folded around a tiny piece of the wire which is then jammed into the holes. The head is rotated so a machine can fill a whole head without moving an inch.

The brushes have got a full head of bristles, but they're a bit rough around the edges, so they set off for a short back and sides. A razor sharp blade cuts away any straggly ends leaving the bristles all the right length. Finally they blunt the bristles' sharp edges.

Across the factory floor the neck of the brush is being assembled. A light sensor tells this machine when a metal pin is in place and it quickly slots on a plastic cover which will hold the head. It's then blasted with rays of UV light. This kills any germs and leaves the plastic completely sterile. A computer scans it, double checks no germs have made it through. next, the necks are given their heads.

So far the casing is hollow but soon a motor and gearbox will put the electric into electric toothbrush. On the left is the gear box and on the right a motor. A machine clamps them together and locks them in place. then a worker welds them altogether with a rechargeable battery. After he checks that the circuit is connected the electrical components are fitted into the casings. A plastic stopper seals the battery in with a simple twist.

The brush heads spin around at a dizzying 40 thousand revs per minute and for good measure they gently rocks back and forth.

A quality controller gives them a final once over. After the necks and heads have been slotted on to the bodies some of the batch go off for testing. They'll be scrubbing false teeth for the next 320 hours – that's well over five years worth of dental hygiene!

If the testers pass then the whole batch is deemed ok and sent out with 5year guarantees. As long as they don't get dropped on the loo, they'll probably be brushing for a lot longer.

Did you know?

4th century Egyptians invented one of the first toothpastes. It was made from salt and pepper, mint and dried iris petals.

牙线

扫描二维码，观看中文视频。

虽然用起来很烦琐，但牙医坚持说，牙线是清洁和保护牙齿的极好方法。它对牙龈也有益处，你不必频繁地去看牙医。

本来牙线是用丝线做的，但现在医药商店的牙线没有那么多情调。现代牙线实际上是由塑料颗粒制成的。首先它们被送入巨大的机器加热熔化。液化塑料像挤出的牙膏一样，变成细而长的线。牙线应该不易断裂。为实现这一目标，新制成的牙线被拉伸。绕过所有线轴和"之"字轨道后，线越拉越快，最终达到每小时 120 千米的速度。拉伸使塑料分子更长和更强。理想的结果是，牙线非常结实，适合剔除牙齿之间的食物屑。

仅此一家工厂每天就生产 1500 万米牙线，没错，1500 万米牙线只用 1 天。充分拉伸后把它绕到大线轴上。这家工厂生产的牙线，足够全部英国人口预防痛苦的牙龈疾病，但研究表明，我们中有近 1/4 的人从来不用牙线。

那些线轴装在这个巨大的机器上。牙线应该很结实，但如果用来清理口腔的角落和缝隙还需要很卫生。首先进行清洗。既能清除污垢也给牙线上一层蜡，有助于在紧密排列的牙齿间通过。接着喷上薄荷粉作为温和的消毒剂。实话说，味道尝起来比塑料线更好。然后把它卷回到大线轴上准备下一个生产工序。

牙线在好的口腔卫生用品中占重要地位，涂蜡的牙线能在口腔中清理牙刷达不到的地方。利用高科技摄像机和测试设备，这个厂的人员不断分析市场上的其他产品，以改进自己的工作。这段测试录像显示，牙刷刚好不能达到代表牙龈的绿线，这对牙齿不利。牙线的好处在于，它能到达牙龈线以下，这点至关重要。牙齿之间的微小缝隙也是难以触及的区域。在牙齿和牙龈之间这些微妙的地方，牙线会起作用。它能清除导致蛀牙和牙龈疾病的食物残渣与细菌。

回到生产线，上百千米的刚做好的牙线要进行包装了。这台机器将缠绕牙线的小线轴分类。安放到位后，转轮把 25 米长的牙线绕到每个轴上。这一过程很迅速，后面还有更多牙线在等着缠绕。接下来是盒子。数百个刚做好的塑料盒放进组装机。把它们排列顺当，准备包装。首先把内支架放进去，接着生产线上的工人插入一卷新的上蜡牙线。

盒子盖住后牙线要电击一下。盒子经受了电击。如果你认为这是一种聪明的消毒办法，那就错了。电弧令氧化外壳，让印着公司标志的油墨层附着在塑料上。

这就是牙线背后的故事，它超细超强，为你预防牙齿糜烂和疾患。

你知道吗？

牙线非常结实，甚至曾在 2002 年被用来越狱。一名囚犯用它划开华盛顿一所监狱的栅栏。

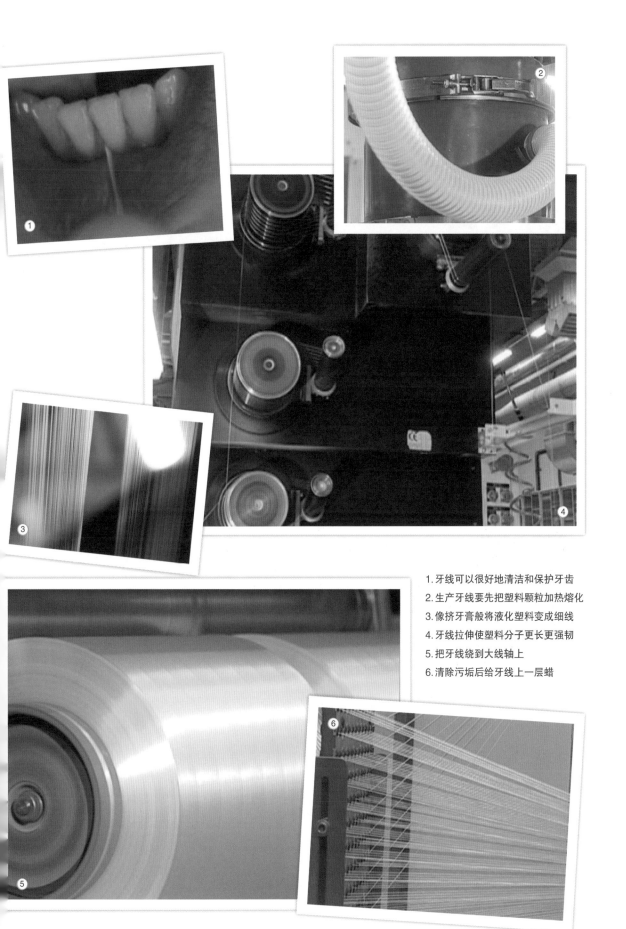

1. 牙线可以很好地清洁和保护牙齿
2. 生产牙线要先把塑料颗粒加热熔化
3. 像挤牙膏般将液化塑料变成细线
4. 牙线拉伸使塑料分子更长更强韧
5. 把牙线绕到大线轴上
6. 清除污垢后给牙线上一层蜡

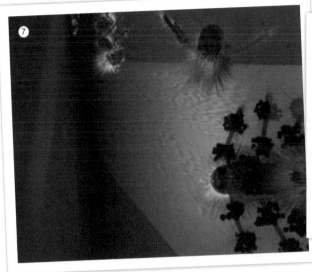

7. 牙线能清除食物残渣与细菌
8. 转轮把牙线绕到每个小线轴上
9. 放进内支架插入一卷新的牙线
10. 电弧氧化外壳
11. 牙线包装完毕准备出厂
12. 牙线使你的牙齿远离疾患

Dental Floss

扫描二维码，观看英文视频。

It's a fiddly job, but dentists insist that flossing is a great way to clean and protect your teeth. It's good for the gums too and hopefully saves you having to visit the dentist too often.

Originally dental floss was made out of silk but the stuff your local chemist stocks these days is a little less exotic. Modern floss is actually made out of plastic beads. Initially they're fed into an enormous heating machine which melts them down. The liquefied plastic is squeezed out like toothpaste into long, thin strands. Dental floss is supposed to be unbreakable. To achieve this, the fresh floss is now stretched. As it passes around all the spools and switchbacks, it's pulled faster and faster eventually reaching 120 kilometres an hour. This stretching makes the plastic molecules longer and stronger. Ideally, the result is super tough unbreakable dental floss perfect for removing food from between your teeth.

This factory alone produces 15 million metres, yes that's 15 million metres of floss every single day. Once it's been adequately stretched, it's put onto these large bobbins. There's enough floss produced in this factory to help keep the whole of Britain free from painful gum disease, but research suggests nearly a quarter of us never floss at all.

For those that do the bobbins are now loaded up onto this enormous machine. The floss may be strong, but if it's going the nooks and crannies of people's mouths it needs to be hygienic too. First, it's washed. This both cleans off any unwanted dirt and coats the floss with a layer of wax which helps it pass between tightly packed teeth. Next it's sprayed with peppermint powder which acts as a mild disinfectant and frankly tastes better than plastic string. Then it's wound back onto the large bobbins, ready for the next stage of the production.

Flossing is an important part of good oral hygiene as the waxy string helps reach many of the parts of your mouth that your toothbrush just can't. Using high tech cameras and test equipment, the staff at this factory continually analyse other products on the market to improve their own. This test footage shows how the toothbrush just can't reach the green line, which represents the gums. This would be bad for your teeth. The benefits of the floss mean they can reach below the gum line which is vital. The tiny gaps between the teeth are also difficult-to-reach areas. It's these tricky places between teeth and gums where dental floss does its work, removing bits of food and bacteria that could lead to decay and gum disease.

Back on the production line, hundreds of kilometres of freshly prepared dental floss are ready for packing. This machine is sorting the tiny spools that the floss will be wrapped around. Once they're in, the rollers get to work to spin 25 meters of floss onto each one. It needs to be a quick process as there are still many more metres of floss to wrap. Next come the cases. Hundreds of freshly moulded plastic boxes are placed in the assembly machine. This puts them the right way up, ready for packing. First they receive an inner holder, and then workers on the production line insert a new roll of waxed floss.

Once the case is closed the floss is in for a bit of a shock. The cases are electrocuted. Now if you think this is a clever way to sterilise you'd be wrong. The electricity oxidises the cases. This helps the layer of ink carrying the company logo to stick on to the plastic.

So that's the unravelled tale of dental floss. The super thin, super strong plastic ribbon that helps prevent disease and decay.

Did you know?

Dental floss is so tough, it was used in a prison break in 2002. An inmate used it to help cut through a fence at a jail in Washington.

牙膏

扫描二维码，观看中文视频。

早晨起床可能不是你最好看的时候，你的呼吸也可能不太好闻。虽然不能解决所有的形象问题，但按常规用牙膏刷牙能使口气清新，并绽开一个美好的微笑。

牙医建议每顿饭后刷牙。虽然我们大多数人不会遵守这条规则，按平均值，你一生中仍然会花 3000 多小时以上来刷牙。那么，一支牙膏里都有些什么东西？我们最喜欢的含氟口味牙膏，从这样的工厂开始生产。

每袋 750 千克的化学配料运到工厂，包括氢氧化铝、碳酸钙和黄原胶。这些听上去很复杂的化学物质只为一个简单目标，保护牙齿不受侵蚀。将每种化学品称出特定的分量，放进大容器里，准备开始制作基本的牙膏。

每个小车上的条形码帮助追踪各个批次。当每种成分充分混合后，将它们送去烧煮。混合物必须在大铝锅中加热，以防止牙膏内发生任何不希望有的化学反应。

现代牙膏有许多好处，包括保持牙齿洁白。显然也有助于防止痛苦的牙龈疾病和蛀牙，它甚至能减轻口臭。回到工厂地面的灶具，混合物已经结合起来形成一种像面团的物质。每批材料制成的牙膏足够填充 30000 管。揭开烧锅后，看起来像一个挂满冰霜的大搅拌碗。用小铲刮掉粘在搅拌叶片上的多余牙膏。

现在该添加一些颜色了。虽然不含任何活性成分，这种混合物能形成著名的彩色条纹。研究表明，比起简单老式的白色牙膏，消费者更喜欢带条纹的品种。彩色液体和单色牙膏混合起来，虽然合成物的绿色看起来不舒服，但完全能让人安全使用。与基础膏体结合之后，最终的混合物看起来不那么刺眼了，它将在挤出牙膏时，构成多色条纹的一部分。

牙膏有不同的颜色和味道序列。最流行和为人熟悉的品种可能是薄荷味。化学家使用香精油，在膏体中加入这种或其他多种风味。

用气相色谱仪将香精油分解成它们的组分。通过这种方式，能在微观层面上看到基本构成。通过组合不同剂量的各种精油，化学家几乎能做出他们想要的任何味道。专门的测试部门每天花几个小时品尝新口味，来评价效果如何。他们还测试工厂的常规口味，如果任何一种有问题，这批牙膏就要被检出。

此时再回到工厂车间，该往管内填充牙膏了，在这里装进的是条纹品种。每种不同颜色通过不同管道进入填装系统。

全世界对牙膏的需求是巨大的，所以这样的工厂有大量工作要做。每台机器每分钟能填充 180 管牙膏。但是生产商怎样让条纹在整管中均匀分布呢？你可以通过这些透明的仿制管子看到，每种颜色在同一时间从不同的喷嘴射出。这样每种条纹都沿管子的长度延伸。

剩下的工作就是把牙膏包装起来，准备装

你知道吗？

在牙膏发明之前，有摩擦性的清洁牙齿替代品包括碎蛋壳和烧焦的动物蹄爪。

进世界各地人们的洗漱袋了。每个箱子都需要称量以确保装满牙膏。24管牙膏应该刚好1350克。没有装足牙膏的箱子达不到这个重量会被立即移除。只有装满的箱子才能被打包、密封和出售。

如果你知道你会在一生中花去3000小时以上站在镜子前刷牙，别以为这是虚度光阴。为了保护和润色你胜利的微笑，花这些时间非常值得。

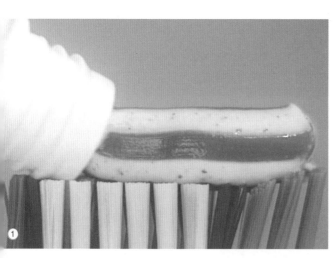

1. 牙医建议每餐饭后刷牙
2. 含氟牙膏的生产从这里开始
3. 称出每种化学品的分量并进行混合
4. 混合原料在容器中加热
5. 用小铲刮掉搅拌叶片上的牙膏
6. 添加颜色让混合物形成彩条

7. 彩色液体和基础膏体混合
8. 各种精油带来不同的味道
9. 测试部门每天品尝新的口味

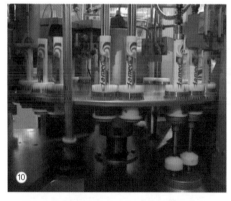

10. 每台机器每分钟填充 180 管牙膏
11. 不同喷嘴挤出不同的色条
12. 牙膏完成包装准备出厂

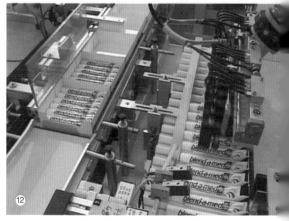

Toothpaste

First thing in the morning, you probably don't look your best, and your breath probably doesn't smell too good either. While it can't solve all cosmetic problems, regular brushing with toothpaste will freshen the breath and bring out the best in a smile.

Dentists recommend brushing after every meal. Now while most of us don't follow that rule every day, on average you will still spend more than 3000 hours of your life brushing your teeth. So what goes into a tube of toothpaste? Production of our favourite flavoured fluoride protection starts in factories like this one.

The chemical ingredients arrive in 750 kilogram sacks and include aluminium hydroxide, calcium carbonate and xanthan gum. Together these complicated-sounding chemicals will work towards one simple aim — to protect teeth from decay. Specific quantities of each chemical are measured into large containers ready to start the process of cooking up the basic toothpaste.

The barcode on each trolley helps keep track of individual batches. Once the ingredients are fully combined, they're sent off to be cooked. The mixture has to be heated in large aluminium pots to prevent any unwanted chemical reactions from occurring inside the toothpaste mix.

Modern toothpaste has many positive benefits including keeping teeth white. Obviously it also helps protect against painful gum diseases and cavities, and it can even alleviate bad breath. Back in the cookers on the factory floor the mixture has combined to form a dough-like substance. Each batch makes enough toothpaste to fill around 30,000 tubes. When a boiler is opened up it looks like a large mixing bowl full of icing on the inside. Spatulas are used to scrape off any excess clinging to the mixing blades.

And now it's time to add some colour. Although it doesn't contain any active ingredients, this mixture will form the famous multi-coloured stripes. Research shows that buyers prefer the stripy variety over the plain old white product. The coloured liquid is mixed with the plain toothpaste, although its synthetic shade of green might look nasty it's perfectly safe for human's to use. After combining with the basic mixture, the final concoction is less garish and will form part of the stripy finish produced by a squeeze on the multi-coloured tubes of toothpaste.

Toothpastes come in a range of colours and flavours. The most popular and familiar variety is probably peppermint. Chemists get this and other flavours into the mixture by using essential oils.

A gas chromatograph is used to break the oils down into their component parts. This way, the basic make-up can be seen at a microscopic level. By combining different quantities of various oils the chemists can recreate almost any flavour they want. A dedicated testing department spends hours each day trying the new flavours to see how they taste. They also test the factory's regular flavours and if any of them aren't right, that batch can be isolated.

Meanwhile back on the factory floor, its time to load up tubes with dental defence, in this case, the stripy variety. Each different colour enters the system through a different pipe.

Worldwide demand for toothpaste is enormous, and so factories like this one have a lot of work to do. Each machine can fill 180 tubes every minute. But how do the manufacturers get the stripes evenly throughout? Well as you can see with these clear dummy tubes, each colour is injected at the same time from a different nozzle. This way each stripe runs along the length of the each tube.

All that remains is to pack them up ready to fill the wash bags of the world. To make sure each box is full they're all weighed. 24 tubes should come to just over 1350 grams. Any boxes that don't contain enough tubes won't make that weight and are immediately removed. Only full boxes are sealed up and sold.

So, if you know you're going to spend over 3000 hours of your life standing in front of a mirror brushing, don't look at it as time down the drain. Think of it as time well spent protecting and polishing that winning smile.

Did you know?

Before the invention of toothpaste, abrasive tooth-cleaning alternatives included broken egg-shells and burnt animal hooves.

牙齿修复

假牙曾经是用木头做的，但今天牙医使用的材料与制造航天飞机隔热层的材料相同。传统的牙冠用覆盖着白色陶瓷的金属制作，但陶瓷可能磨损露出里面的金属。氧化锆是一种不含金属的替代品。这种材料非常强固，用于制造航天飞机的隔热层。

牙医要做的第一件事是麻醉病人。然后使用一台看去像培训宇航员的设备，来测量患者的咬合角度。

如果你害怕牙医，下面请别看——使用似乎所有牙医都偏爱的高功率电锯，切割旧的牙冠并且移除。下面的金属桩也要取出来。

为了安装新牙冠，老的患处必须进行处理。除非你是牙医，否则以前不会见到这个场景。用紫外线光敏树脂做出新的定位点。下一阶段相当的高科技，仅仅为了美观。牙医用激光切除多余的牙龈肉使牙齿保持对称。激光烧掉牙龈肉没有出血或感染的风险，还自带小型抽吸器，吸除所有的碎末残渣。

现在牙医可以制备新牙冠了。首先做出牙齿的石膏模。确保新牙冠成型后能完美的吻合。没谁喜欢让人看出来是戴着假牙。牙医要检查真牙的颜色。让新的假牙颜色一致。在新牙冠准备好之前，病人将戴上临时性的假牙。

模具里倒进石膏，做出实体的牙齿模型，所有缝隙都必须填充。当石膏干燥后，上下牙再次测试，排进新牙冠。感觉异常还是咬合完美，取决于细微的修理调整。

石膏假牙完成后，该做单个牙冠了。牙医以蜡为材料，因为它最便于操作。在许多方面，这项工作更像是雕塑而不是牙科。她用不同色调的蜡来制作模型，看上去尽可能像一颗真牙。这让她更便于观察工作效果。

接下来这位雕刻家将蜡制牙模安装到机器上，这是制作过程的核心部分。她取来一块氧化锆的材料，把蜡模安装到一边，坯料安装到另一边。首先机器会扫描蜡制牙模。控制另一边钻头在什么地方以什么方式切削氧化锆坯料。氧化锆像牙齿珐琅质一样能抵御唾液中的细菌，同样像牙齿珐琅质一样能被钻头轻易切割。

下一步是进入烘箱。在1350摄氏度下让牙冠变硬，但也会大约收缩30%。牙医掌握这一情况，所以制成的牙冠能够与模型完全吻合。

当牙冠做好后你会认为大功告成了。但如果把明亮的白色假牙放进患者嘴里会显得不够自然。该给假牙添点颜色让它和其他牙齿相配。牙医使用粉红色和蓝色的涂料，别以为看起来很可笑，因为还有加热过程。当这两种颜料混合并"烹饪"时便形成了牙齿的颜色。

这样牙医用石膏模型备好新牙冠的位置并做出蜡模，然后用陶瓷和她的颜料盒做出新的假牙。

你知道吗？

在18世纪，治疗牙痛的一种常用办法是把钉子钉进牙龈，直到流血，然后再把钉子钉在树上。

剩下的工作，该让患者试戴了。牙冠被牢牢地安上去，为实现完美的咬合而进行测试，当然最终的检验是一面镜子。理想状态下，患者应该看不出来她的高科技牙冠和自身牙齿之间的差异。

尽管这些由太空材料制成的牙冠再也见不到星星，但肯定能为她咬一口星系巧克力棒。

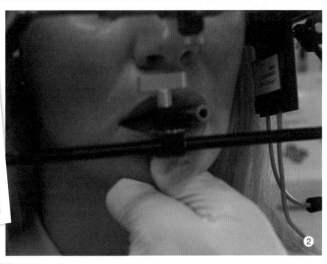

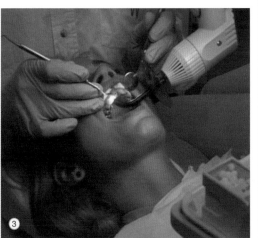

1. 牙医为患者检查牙齿
2. 测量患者的咬合度
3. 用紫外线光敏树脂做出定位点
4. 用激光烧掉多余的牙龈肉
5. 制备新牙冠首先做出石膏模
6. 用石膏做出实体牙齿模型

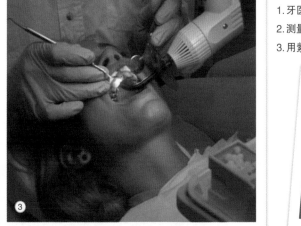

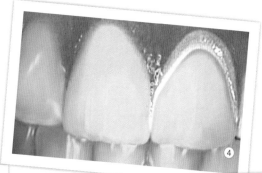

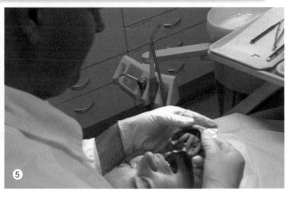

7. 用不同色调的蜡制作牙冠模型
8. 扫描蜡制牙模控制另一端切割氧化锆
9. 把假牙放进烘箱

10. 加热中形成患者牙齿的颜
11. 患者试戴并测试咬合情况
12. 新牙齿让患者心满意足

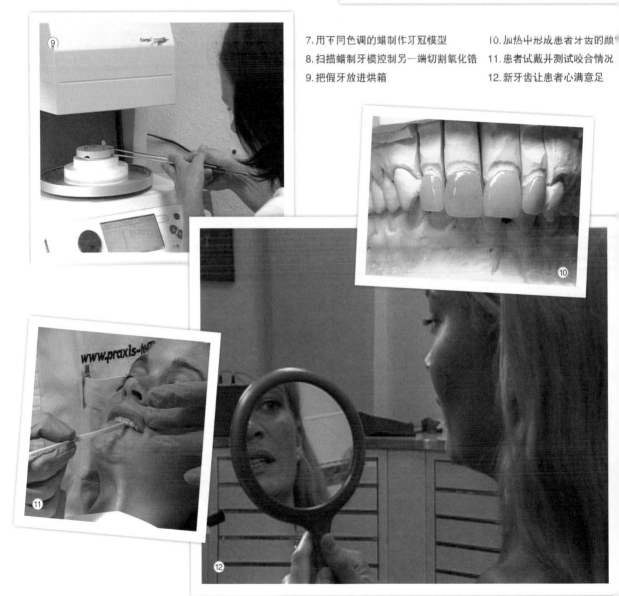

Dental Prosthetics

扫描二维码，观看英文视频。

False teeth may once have been made from wood, but today dentist's are using the same material they use to make heat shields for the space shuttle. Traditional crowns were made of metal covered in a white ceramic, but this covering can wear away revealing the metal inside. However, Zirconium oxide offers a metal-free alternative. It's immensely powerful stuff and has been used to make heat shields for the space shuttle.

First thing the dentist has to do is anaesthetize the patient. He'll then attach a device that looks like something out of an astronaut training programme. It measures the angle of the patient's bite.

If you're afraid of dentists look away now Using that high-powered saw they all seem to love so much, he cuts through the old crown to remove it. The metal peg beneath needs to come out as well.

Then to fit the new crown, the old spaces have to be prepared. You wouldn't have seen this before, unless you were a dentist of course, but using a UV light-sensitive paste, he builds up new anchor points. The next stage is quite high tech but it's only a cosmetic process. Using a laser the dentist will cut away excessive gum flesh to balance out the teeth. The laser cauterizes the flesh so there's no risk of bleeding or infection and it even comes with its own little hoover to suck all the spare bits.

Now the dentist can make the new crowns. First a plaster cast is taken of the teeth. This will help the dentist fashion the new crowns so they fit perfectly. And because no one likes to look like they're wearing false teeth, the dentist also checks the colour of the real teeth. This is so the new one's match. The patient will then be fitted with some temporary implants until her new crowns are ready.

The mould is filled with plaster to create a realistic model of the teeth and the gaps that must be filled. When it's dried, the two sides are tested once again to line up the new crowns. Tiny adjustments are the difference between uncomfortable dentures and the perfect bite.

With the plaster cast in place, it's now time to make each crown. The dentist uses wax because it's easiest to manipulate. In many ways, this kind of work is more like sculpture than dentistry. She'll use different shades of wax to create a model that looks as close as possible to a tooth. This makes it easer for her to see what she's doing.

Next the sculptor will fit it to this machine, and this is heart of the process. Taking a Zirconium Oxide blank, she'll fit the wax model to one side, and the blank to the other. First the machine will scan the model. This tells the drill on the other side where and how to cut the blank down. Like tooth enamel, zirconium oxide will resist the bacteria in saliva, but again, like tooth enamel, the drill can cut through it easily.

The next stage is the oven. At 1,350 degrees Celsius it hardens the crown, but it also shrinks it by about 30%. The dentist knew this was going to happen so what emerges should fit perfectly onto the model.

With the crown completed you would think that would be it. But, if you put that brilliant white replacement into the patient's mouth, it wouldn't look natural. It's time to add the colour so it will match her other teeth. It may look like the dentist is having a laugh using pink and blue paints, but this is because of the heating process. When these two pigments are combined and then cooked, they take on a toothy colour.

So using the plaster cast model, the dentist has worked out the spaces for the new crowns, created wax models, and then made new teeth using ceramics and her paint box.

All that remains now is for the patient to try them out. The crowns are firmly attached and tested for that perfect bite, and of course the final test is the mirror. Ideally the patient shouldn't be able to see any difference between her high tech crowns and her own teeth.

So although these crowns will never see the stars they'll certainly help her nibble on a bar of galaxy.

Did you know?

In the 18th century a common cure for toothache was to drive a nail into the gum until it bled and then hammer the nail into a tree.

隐形眼镜

这里说说隐形眼镜。制造商 1970 年开始生产。在此之前，如果你有近视或远视的话，只能选择眼镜框的颜色和形状。今天许多人却可以选择硬式或者软式隐形眼镜。但它们有什么区别？硬式隐形眼镜是硬的，它的设计完全适合个人的眼睛，所以开始时用角膜曲率仪测量患者的眼睛。虽然眼睛看起来似乎是完全光滑的，在微观层面它们有着不规则的表面形状。实际上你的视网膜和虹膜就像指纹一样，是独一无二的。

制作硬质隐形眼镜从这些塑料棒开始。将它们装入研磨机，患者眼睛的数据被发送到控制计算机里。首先钻石研磨机做出镜片的直径。然后做出镜片内部对应患者眼睛的独特形状，因为眼睛很精细，形状吻合很重要。当形状做好后，镜片被切下。在这个阶段，镜片可以戴在眼睛上了，但却无法看到任何东西。

这家制造商为数百人同时生产隐形眼镜，他们必须确保镜片的正确顺序，不能互相混淆。

镜片粘在一个圆柱上，以便装进研磨机。这台计算机存储了病人眼睛的档案。它控制金刚石钻头两次通过镜片的外侧：第一次达到合适的厚度，第二次做出准确的曲率。镜片做好了，一个很小的镜片却带来视力模糊和目光清晰之间的巨大差异。

取代硬式隐形眼镜的另一种选择是软式隐形眼镜，它会自动适应人们眼睛的形状。对一些用户来说它们更廉价、更方便，因为是一次性的。

软式隐形眼镜的制作从塑料液体开始。蓝色染料被加进去，有助于用户在眼镜掉落时能够找到它们。混合物倒入塑料模具里，顶上加一个盖子，在 97 摄氏度下烘烤 10 分钟。每个镜片都需要检查是否有划痕和裂纹。

在这个阶段它是硬的，每个镜片放进盐水后，会吸收液体变得柔软。再进一步经过热消毒，隐形眼镜就准备好了。

对于那些希望自己的眼睛和手袋颜色相配的人，软性隐形眼镜有多种色调可供选择。镜片装在支架上，放入染料中浸泡。因此，不只是选择镜架的颜色，你现在可以选择你的眼睛颜色了。

你知道吗？

1508 年，列奥纳多·达·芬奇绘制了隐形眼镜的最早版本，不过第一个隐形眼镜直到 1877 年才制造出来。

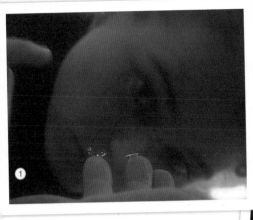

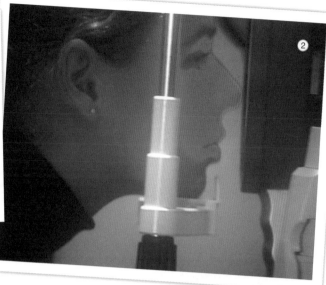

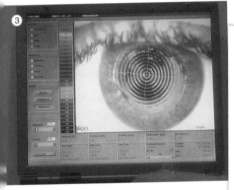

1. 近视可选择硬式或软式隐形眼镜
2. 硬式的要完全适合角膜曲率
3. 人的视网膜和虹膜是独一无二的
4. 将塑料棒装入数控研磨机
5. 钻石研磨机按患者眼睛形状加工
6. 同时生产多个眼镜须确保顺序

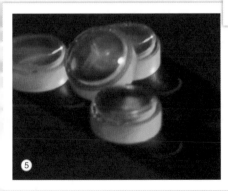

7. 把镜片拧到研磨机上
8. 加工后的镜片使患者视力清晰
9. 软式隐形眼镜自动适应人眼形状
10. 蓝色的塑料液体滴入模具
11. 在模具上加盖烘烤
12. 患者可以选择隐形眼镜的颜色

Contact Lenses

But first, contact lenses. Manufacturers started commercially producing them in the 1970's up until then if you were short or long sighted your choice was pretty much the shape and color of your frames. Now for many a choice of hard or soft lenses. But what's the difference? Hard lenses are rigid and are designed to fit an individual's eye perfectly. So to begin with the patient's eyes are measured with Kerathograph. Though eyes might appear to be perfectly smooth, on a microscopic level they have an irregular surface, in fact you retina and iris is as unique as you finger print.

Hard contact lenses are start out as these plastic sticks. They are loaded into a grinding machine and the patient's eye data is fed into the computer that controls it. First the diamond grinder mills out the lens's diameter. Then the inside is shaped to fit the patients unique pattern, given the delicate nature of the eye this is essential. When it's been shaped the lens is cut off the rod. At this stage the lens would fit on the eye but the wearer wouldn't be able to see anything.

As this manufacturer makes lenses for hundreds of people at a time they have to make sure the lenses are kept in the right order so they don't get mixed up.

The lens is glued to a stud so it can be mounted onto the grinding machine. This computer has the patient's eye prescription stored. It controls the diamond drill which passes over the outside of the lens twice: once to achieve the right thickness, and again for the exact curvature. And there it is: the tiny item that makes the difference between blurred vision and a clear view.

An alternative to hard lenses are soft ones that adapt themselves to the eye. They are cheaper and more convenient for some users as they are disposable.

The start out as a plastic liquid. A blue dye is added to help users find the lenses when they drop them. The mix is poured into plastic moulds, a cover is put on top and it's baked at 97 degrees Celsius for 10 minutes. Each lens needs to be checked for scratches and cracks.

At this stage their rigid, so each one is placed in a saline solution where they absorb the liquid making them soft. Once sterilized with a further blast of heat, they're ready.

And for those who want their eyecolour to match their handbag the soft lenses come in a variety of tints. They're held in a brace and then dipped in a dye. So instead of just choosing the colour of your frames, you can now chose the colour of your eyes too.

Did you know?

Leonardo da Vinci sketched an early version of the contact lens in 1508, though the first one wasn't made until 1877.

创可贴

扫描二维码，观看中文视频。

如果你曾经不小心割伤自己，就会知道创可贴多么有用。这些清洁的具有愈合力的创可贴是如何制作的？制作创可贴的第一步是配制胶水，它是用一种颗粒状橡胶做成。混合过程约需8小时，正确的比例非常重要。如果胶水太强，创可贴可能会给使用者带来过多的伤害。

下一步把大辊帆布装到机器上。它重约一吨，将胶水涂在帆布上，就形成了创可贴的基础。通过这些管子输送胶水混合物，准备添加到帆布上。它看起来像个白色的圆柱体，实际上是由钢球定位的胶水混合物。钢球的重量与滚筒的组合，确保一层薄薄的胶水均匀分布在帆布上。

下一阶段是经过干燥通道。它大约70米长，其中会发生很有趣的过程。

由于伤口需要呼吸，创可贴应该是多孔的。当帆布穿过通道时，细小的气流在胶水层吹出很多小孔。

世上没人会需要1米宽、2000米长的创可贴，所以帆布会被切成便于使用的大小。小刀切入大布卷，生产出只有6厘米宽的创可贴。

但这时创可贴还少一样东西——中间的卫生条。这卷创可贴宽度合适了，工人可以把缺失的卫生条添加到帆布背衬上。卷轴安装到这台机器上，然后慢慢展开。当它通过线轴和之字形路线时，吸水的卫生条被放在中间。这层材料是人造织物，有两个重要功能。首先它吸收伤口渗出的液体，但更重要的是它不会粘在伤口上。放好卫生条后，下一个机器加上两个塑料片，让它在使用前保持干净。一卷创可贴完成后，另一个空卷筒替换它的位置，然后重复整个过程。

当你割伤自己后，开放的伤口很容易感染，创可贴提供了一个临时屏障，在皮肤重新生长时保护伤口。正如我们前面所了解的，创可贴充满了微孔允许空气通过，但不让有害的病原体和细菌进入。现代创可贴含有抗菌能力很强的银微粒。还有另一种辛辣的替代品——信不信由你——是辣椒粉！研究已经表明，辣椒粉的热效应不仅有治疗作用，还会改善血液循环，并能稍微缓解疼痛。

如果所有使用创可贴的人都希望避免一件事，那就是揭掉时拉扯毛发给皮肤带来的刺痛。这家公司不断试验他们的产品以提高性能。他们的目标是生产一种创可贴，需要时能粘得很牢，伤愈后又很容易撕掉。测验时还要监视使用者皮肤的湿度，看创可贴能否让皮肤正常呼吸。

回到生产线上，创可贴该包装了。有家公司声称，打从开始他们已经生产了超过100亿英里长的创可贴，足以绕地球400圈（注：原文如此，计算有误，应为64万圈）。当剪裁好的创可贴沿生产线移动时，机器称出适量的创可贴装盒。分量不对的盒子被气流吹走。只有满盒能够通过。包装完毕，它们被发送给全国的商店和药剂师。

当你下次发生意外，需要把自己贴起来的时候，创可贴已经在恭候你了。

你知道吗？

科学家发明了一种内置微芯片的胶布，可以监测病人的心率、体温和血氧水平。

1. 为创可贴配制胶水
2. 检查胶水混合物的强度
3. 钢球和滚筒使胶水均匀
4. 气流在胶水层吹出很多小孔
5. 这卷创可贴 1 米宽 2 千米长
6. 大卷创可贴被切成便于使用的大小

7. 将卫生条贴在帆布背衬上

8. 用两个塑料片保持使用前清洁

9. 创可贴为伤口提供临时屏障

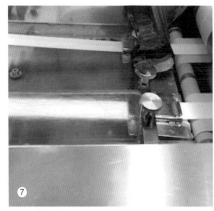

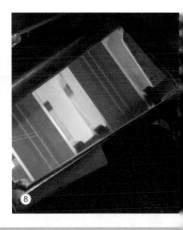

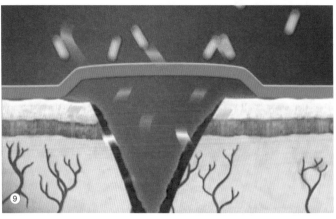

10. 辣椒粉改善血液循环并缓解疼痛

11. 检测撕下时皮肤的刺痛感和透气性

12. 创可贴包装出厂

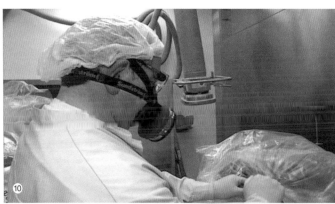

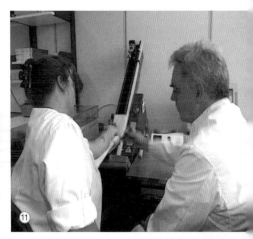

Sticking Plasters

扫描二维码，观看英文视频。

If you've ever cut yourself by accident, you'll know just how useful a plaster can be. But how are these strips of hygienic healing power put together? The first step to making plasters is mixing up some glue. That's made using this granulated rubber. It takes about 8 hours to mix but, it's important to get the proportions right. If the glue is too strong, the plaster might do the wearer more harm than good.

Next large rolls of canvas are loaded up onto this machine. They weigh about a tonne each and the glue will be spread on this cloth to form the basis for the plasters. The glue mix is then pumped through these tubes ready to be added to the cloth. Now it may look like a white cylinder but that is actually the glue mixture being held in place by the steel ball. The weight of the steel ball combined with the rollers guarantees a thin layer of glue, spread evenly across the whole cloth.

The next stage is a trip through the drying tunnel. It's about 70 meters long but a very interesting process occurs inside.

Because wounds need to breathe, plasters need to be porous. As the cloth passes through the tunnel, tiny jets of air blast little holes through the glue.

Now no one in the world could ever need a plaster that's a meter wide and 2 kilometres long, so the cloth is cut down into useful sized pieces. Tiny knives cut into the large roll and produce new plasters just 6 centimetres wide.

But, there's something missing from the plaster at this stage. The hygienic strip in the middle. Now that the rolls are the right width, the workers can add the missing strips to the cloth backing. The rolls are fitted onto this machine and they're slowly unravelled. As they pass along all the spools and switchbacks the absorbent hygienic strip is laid down the middle. This layer is made out of synthetic fleece and has two important functions. Firstly it absorbs any liquid from the wound, but more importantly it doesn't' stick to the wound itself. Once that's been added, the next machine attaches the two plastic strips that keep the padding clean until it's needed. Once a roll of finished plaster strip is complete, an empty spool is put in its place and the whole process is repeated.

When you cut yourself, the open wound is susceptible to infection but the plaster provides a temporary barrier to protect the wound whilst the skin re-grows. As we learnt earlier the plaster is full of tiny pores which allow air to pass through but not harmful germs or bacteria. Modern plasters can include silver particles which have strong anti-bacterial properties, but there's another spicy alternative, believe it or not, it's Chilli powder! Research has shown that the heating effect of chilli powder is not only therapeutic, but improves circulation and provides mild pain relief.

Now if there's one thing most plaster wearers want to avoid, it's the hair-pulling skin-stinging pain of having to pull one off. At this company, they are continually experimenting with their products to improve their performance. Their aim is to produce a plaster that sticks well when needed, but still comes off easily once the wound is healed. Tests also monitor the humidity of the wearer's skin to see if the plaster has allowed the skin to breathe properly or not.

Back on the production line it's time to package the plasters up. One company claims to have produced over 16 billion miles of plasters since they started production. That's enough to circle the globe 400 times. As the trimmed down plasters move along the line a machine weighs out the right amount and boxes them up. Any box without the right amount inside is removed from the line with a blast of air. Only full boxes will pass. Once they're packaged up, they'll be sent out to shops and chemists all over the country.

Ready and waiting for the next time you have an accident and need to stick yourself back together.

Did you know?

Scientists have invented plasters with build-in microchips that can monitor a patient's heart-rate, temperature and oxygen levels.

针头

扫描二维码，观看中文视频。

很少有人喜欢打针，尤其用这么大的针头。

这是一个针头，用来输送药物或给病人打点滴。通常插入手部。因为它们的大小，针头像其他针一样，会引起人们的恐惧——对疼痛的恐惧。然而，大多数人却几乎感觉不到疼痛，因为针头有异常锋利的针尖。

针头由不锈钢制成。工厂把4厘米宽的长条做成圆管。圆管做成后进行焊接封闭。管子的直径大约为1.3厘米，如果你害怕打针，不用担心，没有医生会用它注射。

下面把长管子切成4米长一段。它们用于注射仍然太大了，所以尺寸需要进一步缩小。这个过程被称为牵伸，过大的管子被送进机器，准确地压成适当大小。把一根引导杆送入管内，然后共同插入孔中。

渐渐地大管子受压缩小。强力电动钳将它拉过。这个延伸和收缩管子的过程重复11次之多。最后的结果是细小的针头，让害怕打针的人明显减少畏惧。

然而牵伸过程有一个难题。金属在拉拽中会出现小孔从而减弱强度。解决的办法是把薄管放到1060摄氏度的炉子里。这样能封闭微孔并增加强度。管子做最后一次牵伸之前，只有一根细铁丝能从中穿过并作为向导杆。经过多次牵伸处理，1米长的粗管子能够生产出75米长的针头细管。

75米的针头未免太长，必须切成适当尺寸。由于管子太脆弱，一种能溶解金属的特殊溶液派上了用场。它像一个液体刀片，把细管切成相等长度。如何把这些平头的小管变成锋利的针尖呢，

倘若它们这样脆弱？窍门不是切割，靠研磨。

首先把成批管子固定在一起。第一个砂磨机经过它们磨出一个角度。而针头极其锐利的秘密是顶部额外的缩进。第二次研磨刻出凹陷让针头刺破皮肤的部分更小更尖，并尽量减少痛苦。生产线刚刚做好的新针头看起来很吓人，但正是锋利的针尖能为长期需要注射的患者减缓疼痛和不适。细微的金属碎屑可能在研磨中留在针尖上。这台机器用小玻璃珠击打研磨面，让不平的边缘光滑。

如果针头还有件事必须做到，那就是完全灭菌，所以要进行清洗。2小时的超声波、肥皂和酸足以消毒了。有些针头比普通的皮下注射针头稍粗，用于静脉滴注中把药物送到体内或者采血。较粗的管子使液体能够以更高的速率传输。

这个机器是最后的组装点。针头需要几个零件才能正常使用。针头安上塑料帽。计算机检查是否有针头上下倒插。然后用胶水粘起来，在紫外线下过一遍，使胶水永久固化。最后加上盖子，以利运输安全，并确保针头在使用前无菌。

针头需要检验，测量刺穿人体皮肤需要的力。所需力越大就会引起越多痛苦。庆幸的是对于可能的患者们，这是一种超级锐利又超级高效的针头。

你知道吗？

一家美国公司发明了无针注射法。药物通过贴片上微小的点进入体内。据说是无痛的。

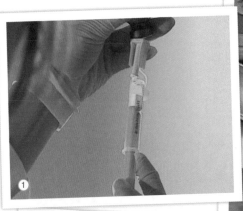

1.针头用来输送药物

2.将长条不锈钢制成圆管

3.长管缝隙焊封后切成 4 米长

4.重复 11 次牵伸使粗管变细

5.薄管经高温封闭微孔增加强度

6.特殊溶液把细管切成相等长度

7. 为小管子研磨出针头

8. 生产线上刚刚研磨好的新针头

9. 用超声波、肥皂和酸进行消毒

10. 针头粘上塑料帽经紫外线固化

11. 加上盖子以保运输安全和无菌

12. 检测刺穿皮肤所需要的力

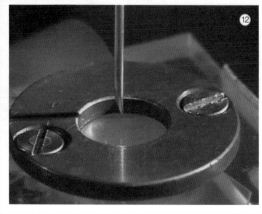

Cannulas

扫描二维码，观看英文视频。

But first, very few people like injections, especially when the needle's as big as this.

This is a cannula, a needle that's used to administer medication or IV drips to patients. They're usually inserted into the hand. Now because of their size, cannulas instil the kind of fear that all needles do — the fear of pain. However, most people find them virtually pain free because of their incredibly sharp tips.

Cannulas are made from stainless steel. The factory uses a band that's 4 centimetres wide, which they shape into a tube. Once it's fully rounded, it's welded closed. This tube has a diameter of almost 1.3 centimetres, but if you're afraid of needles, don't worry. No doctor would use this for an injection.

The next step is to slice this huge pipe into 4 metre sections. They're still way too big to be used for injections, so they need to be reduced in size. In a process called drawing, the over-sized pipe is fed into this machine which literally crushes it down to a better size. A guiding rod is fed into the tube, and the two are inserted into the hole.

Gradually the larger pipe is forced to shrink. A powerful motorised clamp pulls it through. This is repeated up to 11 times lengthening and shrinking the tube. The final result is smaller and significantly less scary for anyone afraid of needles.

However, the drawing process has one setback. Stretching the metal makes it porous which weakens it. The solution is to put the thinner tubes into a furnace which heats them to 1,060 degrees Celsius. This closes any tiny holes and strengthens the steel. By the time the tube is sent for its final stretch, only a piece of thin wire will fit inside as a guide. The multiple trips through the drawing process have made 75 meters of cannula tubing out of just 1 metre of the larger pipe.

Now as a 75 metre needle is a bit long, it must now be cut down to size. Because the tubing is very fragile, a special solution which dissolves metal is used. It acts as a liquid knife slicing the tube into equal lengths. But how do you shape these stubby little tubes into needle sharp needles if they're so fragile? Well the trick, is not to cut them but to grind them down.

First a batch of tubes is attached together. The first sander then carves at an angle across them. However the secret to the super sharp needle is the extra indentations at the top. A second trimming carves these indentations into place which makes the part of the needle that breaks the skin smaller, sharper and hopefully less painful. The finished cannulas might look scary fresh of the production line, but it's these razor sharp points that will minimise pain and discomfort for patients who may need them inserted for long periods. Microscopic slivers of metal may be left on the tip of the needle during the grinding process. This machine blasts the surface with tiny glass beads, which smoothes down any jagged edges.

Now if there's one thing a needle has to be, it's completely sterile, so now they take a bath. 2 hours of ultrasound, soap and acid are enough to sterilise them. Sometimes thicker than ordinary hypodermic needles, the cannula is used to transfer medication into the body through intravenous drips, or to take blood samples out. The thicker tube means more liquid can be transferred at a greater rate.

This machine is the final assembly point. A cannula needs several pieces to work properly. The needles are fitted to plastic caps. And the computer checks the needles haven't been inserted upside down. They're then glued into place and a trip under a UV light solidifies the glue permanently. Finally a cap is added for safety during transportation and to keep the needle sterile until it's needed.

The needles are tested by measuring the force required for the points to pierce human skin. The more force required, the more pain may be caused. Fortunately for potential patients, this is an ultra sharp, ultra-efficient cannula.

Did you know?

A US firm has invented injections without needles. Drugs pass into the body through tiny points on patches. It's said to be painless.

橡胶手套

对于我们之中真正从事保洁工作的人来说，橡胶手套有助于让这项活计稍微变得能够忍受。它们的制造从乳胶开始。天然乳胶通常是白色的，对于清除垃圾时使用的手套，这并非最佳颜色，所以会加入橙色染料。

把人的手浸入乳胶短短几秒钟就做出了一个手套，虽然非常薄，但人不可能整天地站在这里把手泡进乳胶缸中，所以用陶瓷手来充当制造手套的模具。

乳胶不能很好附着在光洁的陶瓷表面，但是将手模在盐水中浸泡并覆盖表面后，乳胶就可以很好地附着了。快速浸蘸一下后，通过转动去掉滴下来的乳胶。

这些手套的厚度已经足够，可以进入下一道工序。但此时还非常脆弱，容易撕裂。这个问题将留在后面解决。

现在需要给手套增加些家庭舒适感——一个棉质衬里。手套制作时是里朝外的，所以当它们浸入这种胶水时，实际上涂层在手套内面。现在，手套有了一个很好的发黏的表面，它将被喷上细棉雾。这层衬里让手套更便于戴上，也感觉更高级，使家务活减少一点烦恼。

要强化乳胶橡胶，需要在炉内进行烘烤。这个过程被称为硫化。这里的两条橡胶可以显示硫化的作用。这条没有经过硫化拉伸后会失去原来的形状，但经过硫化的这条，可以在拉伸后恢复原貌。

手套出炉后就接近完成了，但是它们刚刚被加热到 100 摄氏度，所以还需要在水中冷却。此后工人就可以将它们直接从手模上揭下来了。无须小心翼翼地操作，因为它们是非常结实的手套。在滚筒式烘干机里打个转，他们便最后完成了。

这种精简、高效的工艺流程，让他们每天能做出 5 万双橡胶手套。

你知道吗？

固特异公司 1847 年制造了第一只橡胶手套，用来保护在电报线路上工作的人员免受电击。

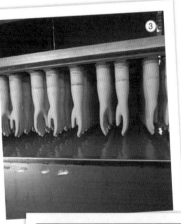

1. 天然乳胶加染料是橡胶手套的原料
2. 工厂用陶瓷手充当手套的模具
3. 手模先在盐水中浸泡
4. 模具在乳胶里快速浸蘸后提起
5. 手套已足够厚可进入下一工序
6. 此时橡胶手套仍很脆弱、易撕裂

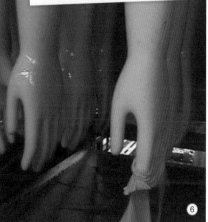

261

7. 浸入胶水后拉起并喷上细棉雾
8. 为增加强度需放进炉内硫化
9. 硫化后的橡胶拉伸后能够复原
10. 快速把手套从手模上揭下来
11. 经过滚筒式烘干机
12. 工厂日产 5 万双手套

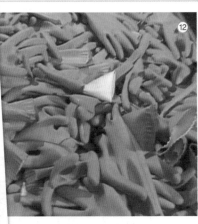

Rubber Gloves

For those of us that acutally do the washing up, rubber gloves help to make the job a little more bearable. To make them, you start off with some latex rubber. natural latex is usually white, which isn't the best color for a pair of golves that do the dirty work, so they add an orange dye.

By dipping his hand into the latex for just a few seconds, this chap has made a glove, albeit a very thin one. But he can't stand around with his hand in the vat all day, so they use ceramic hand to model the gloves instead.

Latex doesn't stick properly to bare ceramic, but a dipping in salt water bath coats the surfaces and then they are ready for the latex. After a quick dip, a spin gets rid of any drips.

These gloves are thick enough to move on to the next step. But right now they are delicate and easily torn. That problem will be sorted out later but now it's time to add a little home comfort. a cotton linning. Gloves are being made inside out, so when they are dipped into this glue, it's actually the inside that's getting coated. Now there's a good sticky surface the gloves go to be sprayed with a fine cotton mist. This lining makes the gloves easier to put on and also gives them a more luxurious feel, making the housework just a little bit less of a chore.

To strengthen the latex rubber, it needs to be baked in a furnace. This process is called vulcanization. These two strips help demonstrate the effect of this process. This one hasn't been vulcanized and loses its shape after being stretched, but the vulcanized strip returns to its original shape.

Once they're out of the furnace the gloves are nearly finished, but they've just been heated to 100 degrees, so they're given a cooling dip. Then the workers can simply pull them off the moulds. There is no need to be delicate with them, these are tough gloves.

A spin in the tumble dryer, and they are finished. This streamlined and efficient process allows them to make 50,000 pairs of rubber gloves every single day.

Did you know?

Goodyear made the first rubber gloves in 1847. They were used to protect the men working on telegraph lines from electric shocks.

天然发刷

大多数人用塑料刷子把打结的头发梳开，这些人显然没听说过全天然毛刷的好处。真正的全天然产品意味着不用塑料，甚至包括手柄。

这些已经干燥了将近一年的榉木板，是手柄的有机替代品。从大木板切下木条，通过刨子使表面平滑，并让厚度一致。接下来木板被切成单个的手柄大小的木块。

方形刷子看上去不对头，拿在手上也不舒服。后面的机器被设计出来解决这一问题。按照下面的轮廓线，上面的磨床把木板雕刻成刷子的形状，包括适合抓握的薄手柄。接着木板被送到滚动容器。所有手柄放进去，容器中装满小石头。小石头的功能是打磨木材表面直到光滑。手柄在加热到45摄氏度的容器中停留大约10小时。加热是为下一个生产工序做准备。

接下来手柄被装在架子上送去浸蜡。木头张开的毛细孔能吸收很多蜡，这有两个好处。首先，吸收的蜡越多就越能凸显木纹，让刷子外观独特。其次，从长远来说，蜡可以保护手柄不被弄脏和损坏。从蜡浴中取出后，手柄放在转轮上晾干。它们不停移动，这样就不会出现蜡滴，影响刷子完成。

现在来讲讲野猪。是的，发刷使用野猪鬃毛，因为它们有助于保持头发健康。

刷毛能把天然油脂从头皮传送到发梢，就像鬃毛在野猪自己身上的作用一样。用理发那样的电推子把鬃毛修剪到统一长度。同时手柄准备好安装刷毛了。

钻孔机在手柄一面钻满成排的孔。鬃毛最终将安放在这里，但它们是如何固定到位呢？这要用到金属丝。你会争论说，金属丝是人造的。但金属是自然的，而人造塑料却不是。一圈金属丝，绞住一撮鬃毛，将它们牢牢固定到孔的底部。

但使用天然猪鬃有什么特别值得称道呢？我们已经知道了鬃毛如何有益于头发养护。还有另一个好处，如这里所展示，塑料刷毛会让头发带上静电。天然猪鬃不容易产生静电，也就不太会把静电传给头发，从而让头发平滑光泽，不会四处胡乱伸。塑料很容易清洁，但天然发刷也一样。用含有少量洗发香波的水洗涤后，猪鬃完好如新。对于美化你的秀发有什么自然选择？出类拔萃的野猪鬃发刷。

你知道吗？

一位丹麦教师发明了一种带有吸尘器附件的发刷，能从头上除掉头虱和虱卵。

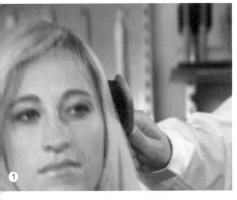

1. 天然发刷对头发有养护作用
2. 手柄由干燥的榉木板做成
3. 锯好刨光的小木块等待成型

4. 磨床把木块加工成刷子的形状
5. 手柄经滚动容器中小石头打磨
6. 手柄在加热后装在架子上

7. 手柄加热持续 10 小时
8. 手柄浸蜡后更耐用并突显木纹
9. 手柄放到转轮上晾干
10. 发刷使用野猪鬃毛
11. 齐整的鬃毛插入刷头小洞
12. 金属丝将鬃毛固定到刷把上

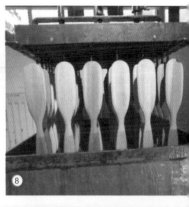

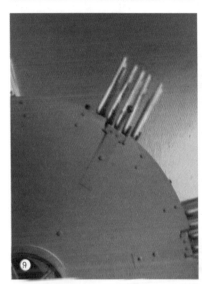

Natural Hairbrush

Most people would use a plastic brush to remove the knots from their hair. But those people obviously haven't heard about the qualities of the all-natural hair brush. A truly all-natural product means no plastic, even for the handles.

So these planks of beech which have been drying for almost a year are an organic alternative. Strips are cut from the big planks and sent through a plane to smooth them down and create a uniform thickness. Next the wooden plank is cut into individual handle-sized pieces.

Now square brushes wouldn't look right, and they also wouldn't sit very comfortably in the hand. The next machine is designed to take care of that problem. Following an outline below, the grinder above carves the wood block into a brush shape including a thinner handle for the grip. Next the wood is sent to this rolling container. The handles are all placed inside, the container which is full of small stones. Their job is to grind down the surface of the wood until its smooth. The handles will spend about 10 hours in the container which is heated to 45 degrees Celsius. The heat helps prepare the wood for the next stage of production.

Next the handles are loaded onto these racks to be sent off for a wax bath. Open pores in the wood allow it to absorb plenty of the wax which has two benefits. First, the more wax the wood absorbs, the more it accentuates the wood grain. This makes the brush look unique. Secondly the wax protects the handles in the long run from dirt and damage. Once they're removed from the bath, the handles are then spun on this carousel to dry. They're kept moving so that drips don't appear which would ruin the brushes' finish.

And now for a touch of wild boar. That's right. Boar bristles are used in these brushes because they help keep hair healthy. By spreading natural oils from the scalp to the hair-tips, the bristles work as they would for the boar itself. The bristles are cut down to a uniform length with what looks like a pair of hair trimmers. Meanwhile the handles are being prepared so the bristles can be added.

A drill is preparing rows of holes all over one side of the handle. This is where the bristles will eventually sit, but how do they get them to stay in place? Well that's where this wire comes in. Now you could argue that wire is man made, but metal is natural where man-made plastics aren't. A loop of the wire is twisted through some bristles and pinned into the bottom of the hole, securing them firmly into place.

But what is the big fuss about the natural bristles then? Well, we've already heard about how they help the hair condition itself. But there's another benefit. Plastic bristles can charge the hair with static electricity, as seen here. However, the natural boar bristles aren't as prone to electricity so they're less likey to pass a charge on to your hair leaving it smooth and shiny and not sticking out all over the place. Plastic may be easy to clean, but the natural brush is too. When washed in water containing a little shampoo, the boar bristles are as good as new. So what's an alternative natural choice for beautifying your barnet? The brilliant boar bristle brush.

Did you know?

A Danish teacher has invented a hairbrush with a vacuum cleaner attachment. It removes head lice and their eggs from infested scalps.

一次性尿布

扫描二维码，观看中文视频。

照料小宝贝，是为人父母的一大乐趣。但抚养孩子中有件事很乏味：换尿布。

现代一次性尿布的发明送来了救星！但这种尿布的生产，比看上去要复杂得多。

木浆纤维是尿布的核心组成部分，你知道后可能会感到惊讶，这种细绒是用松木制作的。除了木绒，工人还加入大量棉花绒。它们在美国种植出来，和松木绒一起制成尿布的软芯。

尿布芯在这台机器中制造。木绒和棉绒在这里混合从一端出来，准备编织成条状。一条长长的混合纤维带出现在另一端，转筒上的铡刀将它切成多个尿布尺寸的小块。

接着添加外层塑料。这个防水层是最后屏障，防止渗漏和污损婴儿衣服及其他物品。然而，现代一次性尿布真正的防护功能，来自这些颗粒状的化学品。这是聚丙烯酸钠，它的超级吸水能力可以防止渗漏。颗粒吸收水分后膨胀变成凝胶将液体锁住。一次性尿布是个大产业，因此各公司不断为市场开发产品。

左侧的尿布没有高吸水性凝胶芯，右边的含有。测验显示出凝胶的双重效果。第一，你能看到右边的凝胶芯尿布吸收液体快得多。第二，"泄漏测试"表明，把干纸巾按压到第一个尿布上吸水部分有明显泄漏。然而当同样的测试在凝胶芯品种尿布上进行时，几乎没有泄漏。化学品起了作用，将所有液体禁锢其中。

超级吸水芯放在我们前面看到的两片柔软木绒之间。再加上一层绒毛织物，由布料制作，类似于戈尔特斯。这将有助于婴儿保持舒适，让皮肤能够呼吸。

所有分层各就各位后，尿布可以剪裁成形。切出大腿部位的缺口，计算机检查每个尿布，确保切割无误。有时切刀失去同步，电脑就能够很快识别有问题的尿布，直接从生产线上取走。

自从20世纪60年代发明出首个原型，现代一次性尿布已经走过了漫长的发展道路，但值得担心的是，每天数以百万计的尿布使用后送到垃圾场，会造成很大的环境问题。公众舆论开始转向可重复使用的尿布，但目前的统计数字显示，90％的英国父母仍然选择更方便的一次性尿布。

生产过程的最后阶段是计数，挤掉空气并包装起来，准备发送到商店。尿布最重要的作用，是将内容物保持其中。然而，宝宝的舒适性也很重要。尿布疹会让婴儿和父母都很难受，但凝胶芯有助于减轻这个问题。

试验表明，皮肤潮湿度降低时，尿布疹的发生率也随之降低。虽然会产生堆积如山的垃圾，大部分父母还是离不开一次性尿布。

你知道吗？

平均来说，一个婴儿一生中会使用5000片一次性尿布。这意味着英国婴儿每天要用掉900万片。

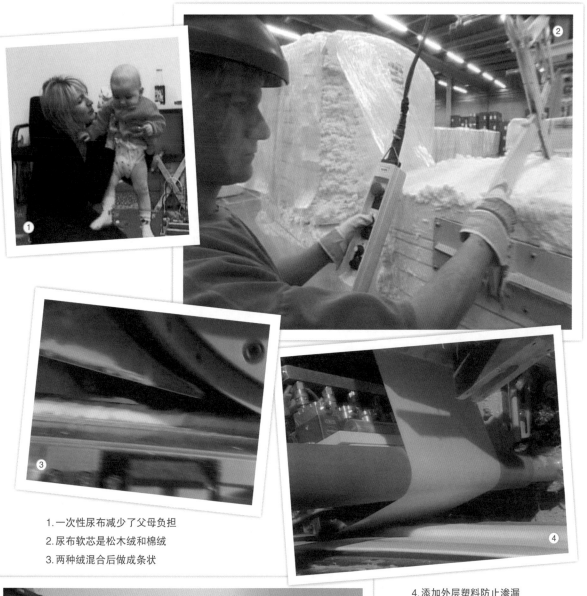

1.一次性尿布减少了父母负担

2.尿布软芯是松木绒和棉绒

3.两种绒混合后做成条状

4.添加外层塑料防止渗漏

5.检测聚丙烯酸钠尿布吸水性能

6.不含凝胶芯的尿布吸水性能差

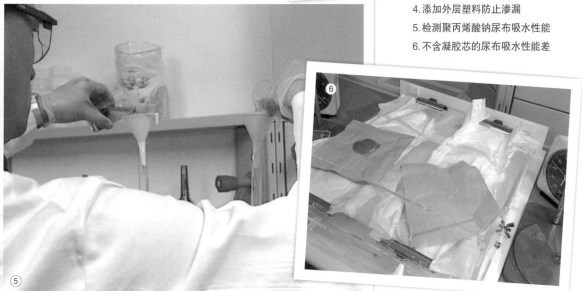

7, 超级吸水芯放在两片末绒间　　10. 生产线上的产品
8　最后在最上层加布料织物　　　11. 挤出尿布中的空气后装袋
9. 分层到位后尿布剪裁成形　　　12. 一次性尿布减少尿布疹发生

Disposable Nappies

扫描二维码，观看英文视频。

The joy of parenthood, that time of life when you have a little one to look after, but there's one part of child-rearing that few relish: nappy changing.

Salvation has come with the invention of the modern disposable nappy, but there's much more to making this innovation than meets the eye.

This pulp will form part of the core of the nappy and you may be surprised to learn that this pulp is made out of pine wood. As well as the wood fluff, these workers are adding gigantic bales of cotton wool. This is grown in the USA and together with the pine fluff will be combined to form the soft core of the new nappy.

The nappy cores are made inside this machine. Here the pulp and cotton wool are mixed together and shot out the other side, ready to be woven into strips. What emerges on the other side is one long strip of pulp, but the guillotine on this revolving drum will cut it into more nappy-sized pieces.

The next addition is the outer plastic. This waterproof layer will act as a final barrier to stop any seepage from spoiling baby's clothes or anything else. However the real protection in the modern disposable nappy comes from this granulated chemical. This is sodium polyacrylate and it's the super-absorbent secret that stops spills. As the granules absorb the water they swell and become a gel which locks the liquid into place. The disposable nappy industry is big business so companies are continually developing these kinds of products for the market.

The nappy on the left here doesn't have a super absorbent gel core. The one on the right does. This test shows the gel's dual benefits. First, as you can see, the gel-cored nappy on the right absorbs the liquid far faster. Second, the "Leakage test" shows that when a dry napkin is applied to the first nappy, there's clear leakage from the absorbent part. However when the same test is conducted on the gel core variety, there's almost no leakage at all. The chemical has done its job and trapped all the liquid inside.

The super absorbent core is layered between two sheets of the soft pulp we saw earlier. Next a fleece will be added. It's made out of a cloth similar to Gore-tex. This will help keep the baby comfortable allowing his or her skin to breathe.

With all the layers in place the nappy can now be shaped. Indentations for the legs are cut out and a computer checks each one to make sure it's been cut in the right place. Sometimes the cutter gets out of sync, but the computer is quick to identify the problem nappies and they are removed from the line straight away.

The modern disposable nappy has come a long way since the first prototypes were invented in the 1960's but there are concerns that the millions used and dumped in landfills every day may create a big environmental problem. Public opinion is beginning to turn back towards the re-usable nappy, but current figures show that 90% of UK parents still prefer the convenient, disposable variety.

The final stage of the production process sees the nappies being counted, squeezed down to remove any air and packed up, ready to be sent to the stores. The most important role of the nappy is to keep its contents inside. However, the baby's comfort is also important. Nappy rash causes grief for both babies and parents but the gel core may help to reduce the problem.

Tests show that when overall skin dampness is reduced, so is the incidence of nappy-rash. While they may be causing a mountain of waste, most parents won't do without disposable nappies.

Did you know?

On average, a baby will use 5,000 disposable nappies during its life. This means UK babies are going through 9 million every single day.

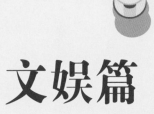

文娱篇

三角钢琴

扫描二维码，观看中文视频。

体形巨大，外观华美，这是一架三角钢琴。这种高雅的乐器，每一架都是由手工精心打造的。三角钢琴与标准立式钢琴的主要区别，在于三角钢琴的琴弦更长，必须水平方向布置。

钢琴的声音，是由绒毡琴锤同时击打三根以上琴弦产生的。琴弦越粗越长，发出音符的基频就越低。每根琴弦的制作，从固定在一台旋转机器上的单根钢丝开始。一位工人用砂纸把钢丝表面磨粗，这样铜线缠绕上去就不会松动。他为琴弦加上另一条铜丝，用来产生更低的基本音符。三角钢琴的琴弦长度从仅仅5厘米到超过2米。因此能产生令人难以置信的宽阔音域。最后工人收紧铜丝的末端，防止琴弦松开。

他们需要使用多层柔软的木片制作钢琴的外壳，让它们能弯曲成传统钢琴的形状。木片被黏合在一起，然后放进压力机中，直至黏合剂干燥。随后木片在另一个压力机中被赋予合适的形状。将它放置到位是一件艰难的工作，需要一点力气和大量的团队协作。

当琴弦被固定后产生总共24吨的拉力。所以钢琴的金属框架必须是超级坚牢结构，由精密铸铁制成，这是19世纪初期便已经发明的技术。青铜粉、溶解剂和增稠剂的混合物被喷涂在金属表面，填充所有的小孔，赋予金属琴架一个光滑的表面。

金属琴架重约170千克，必须用滑轮提升起来，放进木制的琴壳里。

接着就该固定琴弦了，所有的240根琴弦。低音符只有1根琴弦，但中音和高音音符要有2根或3根，从而发音更加洪亮。它们被称为双弦和三弦。每根琴弦必须单独调整张力，直到产生精准的音符。

最后，工人用绒毡琴锤检查琴弦。琴锤则用高速连续发出音符的装置来操控。

钢琴的调试已经就绪，但在安装进外壳中之前，专家还必须检查所有的琴锤是否达到相同的敲击高度，这样才会产生均匀的声音。

琴键必须在受到52克重量时下沉，琴键要一个接一个进行测试。谁说音乐工业只有风光无限？

剩下的活就是最后组装。每一架三角钢琴都是以最高标准手工制作的，因此，这家工厂平均每天只能生产一架钢琴。✎

你知道吗？

有史以来最贵的钢琴是约翰·列侬用来创作《想象》的那台，售价167万英镑。

1. 高雅的三角钢琴由手工打造
2. 琴弦长且水平布置是它的特点
3. 将制作琴弦的单根钢丝打磨粗糙
4. 为琴弦绕一条铜丝产生更低音符
5. 黏合的木片放进压力机中
6. 钢琴外壳成型

7. 滑轮把铸铁琴架放入木制琴壳
8. 240 根琴弦必须单独调整张力
9. 用绒毡琴锤检查琴弦
10. 琴锤必须达到相同的敲击高度
11. 琴键在受到 52 克重量时下沉
12. 钢琴组装完成

Grand Piano

It's big, it's beautiful, it's a grand piano. And each one of these majestic instruments is exquisitely handcrafted. The main difference from a standard upright piano is that in a Grand the strings are longer and have to be arranged horizontally.

The sound is produced by a felt hammer hitting up to three strings at once. The thicker and longer the strings are, the lower the notes. Each string begins life as a single steel wire attached to a spinning machine. One of the workers scuffs up the surface with sand paper, so when a copper wire is wrapped around it will stay in place. He adds another wire to the string to give it a deeper base note. The strings for a grand piano range in the length from just five centimetres to over two metres. That's what allows them to produce an incredibly wide range of notes. Finally he tightens the end of the wire to stop it from uncoiling.

They have to use layers of flexible wood for the piano's casing so they will be able to bend it into the traditional shape. They are glued together and then squashed in a press until the adhesives dry. Then it's taken over to another press which will give it it's shape. It's a tough job to get it into position and requires a bit of muscle and lots of team work.

When the strings are fixed they will have a combined tension of twenty four tons. So the metal frame needs to be a super tough structure. It's made from precision cast iron, in a technique that was developed in the early nineteenth century. A cocktail of bronze powder, dissolver and thickener is sprayed onto the metal surface to fill any pores and give the frame a smooth finish.

The metal frame weighs about 170 kilograms, and has to be lifted into the wooden casing with a pulley.

Then it's time to fix the strings: all 240 of them. Low notes have just one string but the middle and higher notes have two or three to make them louder. They are called bichords and trichords. Each individual string has to be stressed until it reaches the exact pitch.

Finally they check the strings with a felt hammer. The hammer's controlled by a mechanism that enables a series of notes to be played consecutively at high speed.

The workings of the piano are now complete but before it can be fitted into the casing an expert must check that all the hammers reach the same keystroke height so they produce an even sound.

A key has to drop under a weight of fifty two grams and they all have to be tested one by one. Who said the music industry was glamorous?

All that remains is the final assembly. Every individual grand piano is hand crafted to the highest standards, that's why this factory builds on average just one a day.

Did you know?

The most expensive piano ever sold was used by John Lennon to write Imagine and was sold for £1.67 million.

电吉他

扫描二维码，观看中文视频。

滚石乐队的歌词唱道，"这是唯一的摇滚乐，但我喜欢它"。可如果没有电吉他，哪有滚石乐队、甲壳虫乐队或其他的当代摇滚明星呢？音乐界这一革命性的进展，因吉米·亨德里克斯和埃里克·克莱普顿这样的歌星而闻名。电吉他与原声吉他同根同源，都是用木头制作的。

用于做面板的枫木必须干燥，否则会变形，让吉他发出难听的声音。这个干燥过程大约需要2年时间，但技术已经进步了，如今木材在这个60摄氏度的特殊烘箱中只花6周时间就能得到同样的效果。

首先将厚枫木切成两半。天然木纹将提升吉他的美观，两块木板粘在一起时形成对称图案。这两块木头粘起来，在压力下放置约一小时，木头便永久粘牢了。

待到面板晾干时，另一名工人可以开始制作琴颈了。这把吉他用的是非洲紫檀，一种原产于西非的硬木。一根金属条用来加强琴颈。琴弦产生80千克的拉力，这根金属条能帮助木头承受负荷。接下来工人加上一个乌木指板。乌木也是硬木，它很结实，足以承受弹奏者的手指摩擦。

同时制作琴身的工人正在打造吉他的背面。它通常用桃花心木制成，具有良好的自然声音共振性能。工人用机械车床钻出拾音器的位置，然后把枫木面板和琴身粘在一起。琴身再次放到车床上。吉他正面可以加工成形了并且雕刻出安放电子元件的位置。随着钻头快速更换，琴身的其他部分也相继完成。

电吉他的传统形状逐渐显现。最后的任务需要技师用手工完成。木匠打磨琴身凸现出木头的自然纹理和图案。接下来改善吉他外观的另一个手法是染色。染料之后是清漆涂层。它能保护吉他免受基本损害，但遇到一位把自己当作皮特·汤森的吉他手就无能为力了。

当这一切都在进行时，琴颈和指板已经制作完毕。加上指板贴，有助于弹奏者在学习时知道手指放在何处。这是琴衍，激光引导机器人把它牢牢压进木头里。

吉他手将琴弦按到琴衍上，发出不同的音符，因此必须把琴衍放到精确位置。

琴枕用胶水固定到位，琴弦靠琴枕在指板上撑起。最后琴颈用蜡抛光。它类似于家具蜡，能保护琴颈，同时有助于弹奏者的手上下平稳滑动。

下面该安装电子元件了。原声吉他用一个大的空腔把琴弦振动声音放大。电吉他用磁铁来"拾取"声音。当琴颈和琴身结合在一起后，就可以插入电磁铁了。它们能把吉他手的弹拨变成声音。

剩下的工作就是安上用钢丝绕制的琴弦。不同规格的琴弦产生不同的音符。

你知道吗？

在卡塔尔的一场慈善拍卖会上，由米克·贾格尔和埃里克·克莱普顿等明星签名的一把芬达吉他以157万英镑的价格成交。

它们需要拧紧到适当的张力，这靠琴颈顶部的弦钮来完成。多余的琴弦被裁掉。

如我们所知，电吉他的拾音器是电磁铁。当琴弦拨动时，它们和磁铁之间的距离发生改变。这种变化以电子信号方式被纪录，并传递

到放大器。

来自放大器的信号送到扬声器，至于发出的是噪声还是音乐，就取决于你是否喜欢演奏者了。女士们，先生们，这就是世界著名的电吉他！✎

1.摇滚乐队离不开电吉他
2.面板用枫木，干燥后锯成两半粘牢
3.琴颈用非洲紫檀并用金属条加强
4.背面钻出拾音器位置
5.吉他成形并雕出电子元件位置
6.琴身经手工打磨后涂色上清漆

7. 琴颈和指板完工后加指板贴
8. 激光引导下将琴衍压进木中
9. 琴枕用胶水固定在指板上
10. 将琴颈和琴身安装在一起
11. 为吉他安装琴弦
12. 电吉他的声音送进扬声器

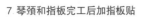

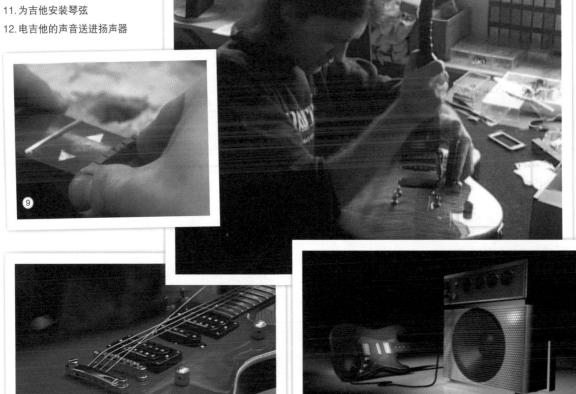

Electric Guitars

According to the Rolling Stones lyrics, its only Rock and Roll, but I like it. But where would The Stones, The Beatles or any modern rock star be today, without the electric guitar? This revolutionary musical advance was made famous by stars like Jimi Hendrix and Eric Clapton. But the electric guitar's roots are the same as the acoustic variety. They're made of wood.

The maple used here for the front has to be dried first, otherwise it would buckle and the guitar would sound awful. The process used to take about 2 years, but technology has moved on. The wood will now spend just 6 weeks in this special kiln at 60 degrees Celsius and get the same result.

First a thick plank of the maple is sliced in half. The natural wood grain will enhance the guitars appearance when the two sides are stuck together creating a symmetrical pattern. These two halves are glued together, and left in a press for about an hour which should bond the wood permanently.

Whilst the face is drying another worker can start building the neck. On this guitar the worker is using Ovangkol, a hard wood native to West Africa. A metal bar is used to reinforce the guitar's neck. The strings exert 80 kilos of force and this bar helps strengthen the wood to withstand the pressure. Next the guitar maker will add an ebony fingerboard. Ebony is also a hardwood that's strong enough to survive the friction from the guitarist's fingers.

Meanwhile the body maker is constructing the back. This is often made from mahogany which has excellent natural acoustic resonance. Using a robotic lathe, he drills out the positions for the pickups and then glues the maple face to the body. The guitar body is once again fitted into the lathe. The front can now be shaped and positions for the electronics are carved out. With a quick change of the drill head, the rest of the body can now be formed.

The traditional shape of an electric guitar is emerging. The final work requires the hand of a craftsman. A carpenter sands the body dow to bring out the natural textures and patterns. Another factor that will improve the appearance of the guitar is the dye which is added next. This is then followed by a layer of varnish. It'll protect the guitar from basic damage, but not from a guitarist who thinks he's Pete Townshend.

Whilst all of this is going on, the neck and fret board are being finished off. Fret markers are being added in. They help guitarists to know where to place their fingers when they're learning. And these are the frets. The laser guides the robot and they are pressed firmly into the wood.

The guitarists will press the strings into the frets to make the different notes, so it's important they're accurately lodged in place.

The strings are held above the fretboard by the nut which is glued into place next. Finally the neck is polished with wax. It's similar to furniture wax and protects the neck whilst helping the guitarist's hand to slide smoothly up and down.

The next stage is the electronics. Acoustic guitars use a large air space to amplify sound made by the vibrating strings. Electric guitars use magnets to "pick up" the sound. With the neck and body now joined together, the electro-magnets can be inserted. They will help turn the guitarist's strumming into sound.

All that's left is to add the strings which are made from wound steel. Different gauges string produce different notes. They need to be tightened to the right tension which is done using the tuning pegs at the top of the neck. Spare string is removed.

As we've already learnt, an electric guitar's pick ups are electro magnets. When the strings are strummed they change the distance between themselves and the magnet. This change is registered electronically and passed on to the amplifier.

From the amplifier the signals are passed on to the speakers which reproduce the noise, or music depending on whether you like the musician or not.

Ladies and gentlemen, the world famous electric guitar!

Did you know?

A Fender Stratocaster signed by star such as Mick Jagger and Eric Clapton sold for £1.57 million at a charity auction in Qatar.

琴弦

扫描二维码，观看中文视频。

一把高质量的经典小提琴，是件美丽的乐器，它的精湛工艺受到天才音乐家的称赞。但是你知道吗？为了做出专业的琴弦，还需要屠宰高手和小羊羔吗？

是的。再现莫扎特最优秀作品的琴弦是用羊肠做的。塑料替代品发不出同样的声音共振，因此要得到真正的原声，非羊肠线不可。

在欧洲每天有成千上万只羔羊被宰杀做食物。羊肠虽不经常出现在菜单上，却没有浪费掉，而在经典音乐世界得到很好的利用。

制作琴弦的最好、最柔韧的肠子，都来自不超过 4 个月大的羊羔。但使用的并不是肠子本身。琴弦制造商想要获得的纤维组织叫胶原蛋白。它从羊肠提取，然后清洗和腌制，杀灭可能存活的细菌。但不能在这里待得太久，否则盐会使它脱水。多余的盐分被冲干净，留下一堆绳索样东西，看起来像意大利面条。

然后肠组织纤维经过化学处理。加入过氧化氢用来漂白纤维，再用织物柔顺剂使它保持天然弹性。经过再一次彻底冲洗后，纤维组织就可以进行分类和分级了。

取自幼小稚嫩的羊羔，材料具有顶级的质量，但只有其中最好的才能用来制作乐器的琴弦。需要训练和经验，才能识别高品质的纤维，让它在音乐厅产生最好的声音。

当最好的纤维选定后，一个分离机将纤维分成小股。在纤维管顶部划开微小的切口，然后送到一端安有刀刃的金属片上。胶原纤维带被切割成不同厚度的长条，较厚的纤维条将被用于大提琴或竖琴这样的大型乐器。较薄的纤维条留给更精密的乐器，比如小提琴。

但这些纤维还不能演奏。它们必须先结合起来。音符越低，琴弦就要越粗。例如音符"E"，由 5 条纤维组成，"A"需要 7 条，"D"由 10 条做成。

琴弦制造商把纤维缠在一起，把它们连接到一段打结的普通绳子上。然后这些绳圈被装上一个绞拧机，把纤维拧在一起。这样会挤出多余的水分，使纤维拧紧并粘在一起。然后，新做成的琴弦被挂起来干燥 4 天。琴弦制造商密切关注干燥过程，以确保没有扭结、卷曲或散开的纤维。

当干燥完成后，他把琴弦从支架上剪下来，带它们进行最后加工。因为每根琴弦是多条纤维的组合，所以非常粗糙，将使演奏非常困难。下一步是让每条琴弦变得平滑。用激光导引器找到每条琴弦最细的部分，然后用研磨机把整条弦减缩到这个统一宽度，磨光任何粗糙的边缘。这样，当初粗糙的纤维组合，现在变成了一条长长的、连续的音乐级别肠线，可以用在经典小提琴上。

还有最后一件事要做。这些纤维对小提琴还嫌太长，所以生产过程的最后一步是把它修剪到合适长度。然而，在这里不需要复杂的设备。尺子和剪刀就够了。琴弦制造商把每束琴弦量出 60 厘米，剪到合适长度。

剩下的事，把每根琴弦各自包好。琴弦可以提供给最苛刻的音乐家使用了，他们寻求只有天然肠线才能发出的特别音质。所以演奏真正的经典音乐需要肠子。从协奏曲到烤肉串，人们都要感谢厥功甚伟的羔羊。

1. 演奏莫扎特作品的琴弦用羊肠做成
2. 专业琴弦需要屠宰高手和小羊羔
3. 羊肠被清洗和腌制以杀灭细菌
4. 漂白肠纤维加入织物柔顺剂
5. 最好的材料才能制作琴弦
6. 切割胶原纤维管

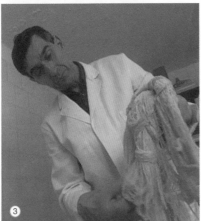

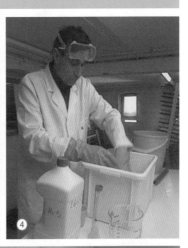

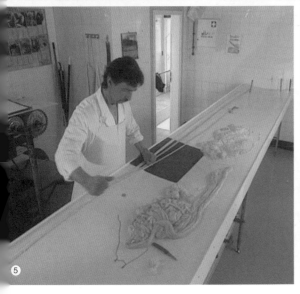

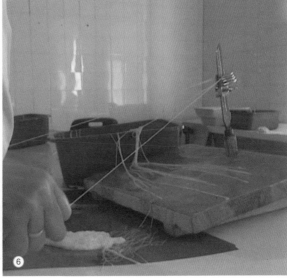

7. 较厚的纤维条用于大型乐器

8. 音调越低琴弦就越粗

9. 绞拧机把纤维水分拧出

10. 干燥后把弦研磨到同样宽度

11. 按尺寸剪开并进行包装

12. 音乐家永远离不开琴弦

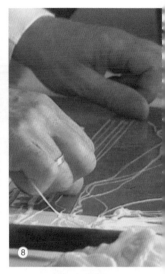

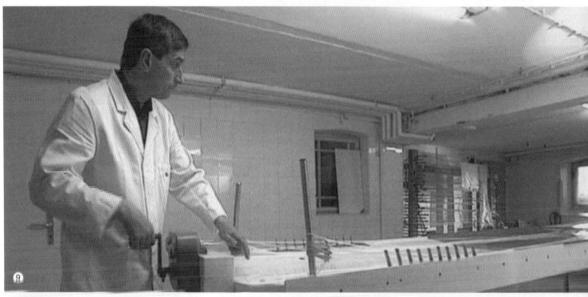

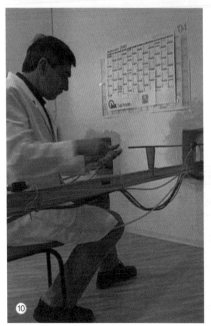

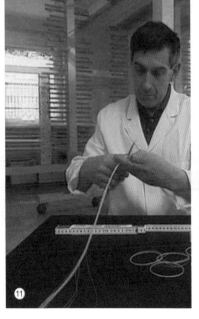

Instrument Strings

A quality classical violin is a beautiful instrument, its craftsmanship complimented by the talented expert musicians. But did you know that to make specialist strings you need an expert butcher, and a lamb?

That's right. The strings used to reproduce some of Mozart's finest compositions are made from lamb's guts. Plastic alternatives don't give the same acoustic resonance so for that authentic original tone, only sheer guts will do.

Everyday throughout Europe many thousands of lambs are slaughtered for food. But rather than waste the guts which are not often found on menus, they're put to do good use in the world of classical music. The best, most supple intestines for making strings come from lambs that are no more than 4 months old. But it isn't the guts themselves that are used. The string makers are trying to extract fibrous tissue called collagen. It's retrieved from the guts, then cleaned and salted – this kills off any bacteria that may be present. But they can't spend too in here or the salt would dehydrate them. The excess salt is rinsed off, leaving a stringy pile that looks like spaghetti.

The intestine fibres are then chemically treated. Hydrogen peroxide is added to bleach the fibres … and fabric softener is used to help them to retain their natural elasticity. After another good rinsing the fibrous tissue is ready to be sorted and graded.

Sourced from young supple sheep this material should be of tip top quality, but only the best of the batch will be deemed good enough to make strings for musical instruments. It takes training and experience to identify the quality fibres that will produce the best sound in a concert hall.

When the best parts have been selected a separating machine is used to split the fibres into smaller strands. Tiny cuts are made at the top of the fibrous tubes before they're fed over a piece of metal with blade at the end. Bands of varying thickness are then cut from the collagen strips. Thicker bands will be used for the bigger instruments like the Cello or the Harp. The thinner bands will be reserved for the more delicate instruments such as the violin.

But the fibres aren't ready to be played just yet. They must be combined first. The deeper the note, the thicker the string needs to be. An "E" for example is made up of 5 fibres, an "A" needs 7 and a "D" is made out of ten.

The string maker winds the fibres together, and at-

taches them to a knotted piece of ordinary string. These string loops can then be attached to a wringer, which spins the fibres together. This squeezes out any excess water and twists them tightly so they bond. The newly fashioned strings are then hung up and left to dry for 4 days. The string maker keeps a close eye on the drying process to ensure there are no kinks, warps or loose fibres.

When they're done he cuts the strings from their holders and takes them to be finished. Because each string is a combination of fibres, they're very rough, and this would make playing very difficult. The next step is to smooth each string down. Using a laser guide that finds the thinnest part of each string, a grinder then reduces the whole length to this same uniform width, smoothing any rough edges as it goes. So what started out as a rough fibrous combination, is now one long continuous piece of musical-grade gut ready to be used on a classical violin.

There's just one last thing to do. These fibres are still far too long for a violin, so the final stage of the production process is to cut them down to the right length. However, no sophisticated equipment is needed here. A ruler and a pair of scissors are fine. The string maker will measure out 60 centimetres and cut each bundle down to size.

All that remains is to pack the individual strings into their respective packets, ready for use by the most demanding musicians looking for that special acoustic sound quality that's only produced by natural gut. So playing truly classical music literally takes guts and from a concerto to a kebab many people owe thanks to the versatile young lamb.

乐高玩具

扫描二维码,观看中文视频。

传统老玩具无疑是最好的。乐高是如此受到欢迎,尽管 50 年前才被发明出来,它已经两次被选为 21 世纪最佳玩具!大部分乐高积木,诞生于丹麦这家工厂的细小塑料颗粒。14 个筒仓装满不同颜色的塑料颗粒,共有 33 吨。

塑料颗粒从管道中吸出后被熔化,成形为乐高砖块,能够搭建成宇宙飞船、城堡和农场动物。

这个大厅里有 1000 多台机器,昼夜不停地生产着乐高玩具砖块。但却没有一个工人。这些机器在 232 摄氏度的高温下把塑料颗粒熔化,并对它们塑形,冷却。然后从机器出口掉落到下方的容器里。做完所有这一切,只需要不到 10 秒。这样的快速流程,让乐高公司每小时生产 170 万块塑料玩具积木。累计起来,乐高公司一年能生产数量庞大的 150 亿块。

工厂几乎完全自动化。当一个容器装满后,自动车会取走容器,把它送到储存大厅。在宽阔的厂房里,另一个自动化系统把每个箱子放到特定架子上。

但乐高并不仅仅是一堆塑料块。在乐高世界里,还有各种各样微小的人物。

这台机器每小时生产出 15000 个玩偶脑袋。橡皮图章给每个头都印上一张脸。有笑容可掬的好孩子,也有脸色阴沉的坏家伙。小人儿们身体的制作描绘同头部的流程一样。这里是玩偶小人身首拼接的地方,消防队员将得到他们的制服,骑士会穿上盔甲,海盗则被装上塑料的"木腿"。

你现在也许还认不出这个著名的角色,但很快就会真相大白。当摄像头在生产线上发现小人儿的校服穿反了时,会用一股气流将它吹出去,从头再通过一次。当小人儿的头和身体连起来后,一个乐高版的哈利·波特出现了。

在工厂的另外地方,一个设计师正在工作,构想着搭建乐高玩具的新花样。研究人员已经发现,仅用 6 块常规的乐高积木就有超过 9 亿种组合。所以这位设计师是没有理由交白卷的。比起乐高乐园的巨型雕塑,这些盒装的乐高玩具就显得微不足道了。120 个全职的乐高设计师在仓库里绞尽脑汁,用他们的"乐高头脑"创造一切能想到的东西。

如果你觉得金字塔很难修造,不妨设想用这些大型乐高雕塑搭建一座。即使有计算机指导,这仍然是个复杂而耗时的差事。仅一个模型就用掉几千块乐高积木,需要长达一个月的时间去完成,乐高以简单而闻名。但用它做出的一些雕塑却十分复杂。乐高的多功能性是它永葆魅力的原因。这使得乐高公司在全球销售积木 2000 亿块之多。

你知道吗?

乐高这个词来自丹麦语"leg godt",意思是"玩得好"。

1. 乐高被发明 50 年来两次当选世纪最佳玩具
2. 筒仓装满不同颜色的塑料颗粒
3. 塑料颗粒在机器里熔化冷却
4. 装满乐高积木块的箱子被放到架上
5. 乐高玩具世界还有一些小人儿
6. 穿上校服、披上盔甲的哈利·波特

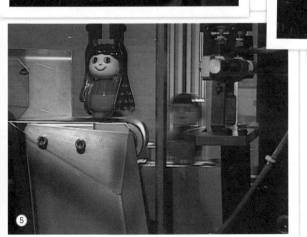

7. 设计师构想乐高玩具新花样

8. 用计算机设计出能想到的东西

9. 已经完成的巨轮与河马

10. 简单的乐高能做出复杂雕塑

11. 摆动大脑袋的犀牛充满魅力

12. 乐高造出的人物栩栩如生

Lego

But first…lego, the old toys are definitely the best. And Lego is so popular that despite only having been invented 50 years ago it's been voted toy of the century twice! Most Lego begins life as tiny granules of plastic at this factory in Denmark. There are 14 silos full of different colours containing 33 tons in total.

Pipes suck the granules out of the silos on their way to be melted down, moulded into bricks and then built into spaceships, castles and farmyard animals.

In this hall there are over a thousand machines making the bricks 24 hours a day, but there's not a single person in sight. The machines melt the granules at 232 degrees Celsius, shape them, cool them down and then spit them out into the containers below all in under ten seconds. This rapid process helps Lego to make 1.7 million bricks every hour. That adds up to an enormous 15 billion a year.

The factory is almost completely automated. When a container is full, a robotic car collects it and takes it to the storage hall. In a vast hall another automated system takes each crate to its specified shelf.

But Lego isn't just a bunch of plastic bricks the world of Lego is populated with all kinds of tiny people.

This machine spits out 15 thousand heads an hour. A rubber stamp gives the heads a face. Smiles for the good guys. Scowls for the baddies. The bodies are made and painted just like the heads, and this is where they'll be joined together. Firemen will get their uniforms, knights don their armour and pirates will be given their plastic wooden legs.

You might not recognise this well known character at the moment but all will be revealed! When a camera spots a school uniform the wrong way round the body is ejected by a blast of air to come round again. As the heads and bodies are attached a Lego Harry Potter appears.

In another part of the factory a designer has got the job of thinking up yet another new way to put the bricks together. The boffins have worked out that with just 6 regular bricks there are over 900 million combinations so he's got no excuses for drawing a blank. But box sets are small potatoes compared to the huge sculptures in Lego land. A hundred and 20 full time designers work in warehouses to create everything and anything they can conjure up in their Lego-driven minds.

If you thought that a pyramid was tricky to build, just imagine trying to make one of these. Even with computers to guide them it's a complicated and time consuming business. With thousands of bricks being used with each model it can take up to a month to build one. Lego has made its name by being simple but some of the sculptures it can be used to create are anything but. Its versatility is its key to its enduring appeal and it's helped Lego sell an amazing 200 billion bricks worldwide.

Did you know?

The word "Lego" comes from the Danish phrase "leg godt" which means "play well".

扫描二维码，观看中文视频。

激流勇进

"激流勇进"是博览会最受欢迎的游乐设施之一。建造起来却无疑是艰辛的工作，负责工程的人可以告诉你其中甘苦，因为博览会要从一个城镇流动到另一个城镇，所以他们几乎每个月都要建造一次。

施工需要大约一个星期，第一个阶段是备好基础。显然，对于"激流勇进"来说，基础必须防水。他们正在建造世界上最大的可移动"激流勇进"。注入超过 150 万升的水。

"激流勇进"一年内要搭建 10 次以上，难免导致磨损，所以必须定期维护设备。基础上的裂隙用硅胶密封剂修复。把框架插到这些孔里，橡胶圈形成防水密封环。用起重机把构件吊过来，工作人员引导它们安装到位。所有激流勇进的结构用火车运到，一块块地卸下。

当顶部的框架仍在建设中时，下面第一节水槽开始安装。工程全部完成后，总共有 52 节水槽。这些钢梁将引导小车沿着水槽滑行。20 人的团队把整个"激流勇进"搭建起来。一天结束时，他们会取得显著的进展。

第一天早上，他们先固定上部分的承重横梁。这些人为好玩的博览会工作，却有马戏团杂技演员的本领，因为需要在高达 26 米的横梁上保持平衡。

工作突飞猛进，时间转瞬即逝。最后的一段水槽安装到位，整个水道宣告完成。他们用起重机把"激流勇进"的小车吊进来。每个价值超过 5000 英镑。

实现水上乘坐的下一个问题是水。把整个系统充满，需要 150 万升水，花费将近 48 小时，耗资 5000 英镑。

到第 5 天结束时"激流勇进"建成并开始运行了。但在公众乘坐之前还有很多事情要做。

空车被送去运行一天，以确保"激流勇进"的所有连接没有松动，并且 5000 个零件全部按照标准正常运行。系统出现任何问题，都将在控制室中显示，机械师会检查清楚。

传送带必须足够结实，能同时承受小车和 6 个乘客的重量。传送带把游客从底部运到 26 米的高度。这未必是乘客永生难忘的旅程，但却能提供轻松的娱乐。

系统周围放置着 10 个大水泵，用来保持水的压力和调节流量。这些水泵产生巨大的压力，让整个系统每小时流过的水量近 30000 升。

测试结束，系统安全可靠，工人们终于可以先娱乐一下了。最好在可能的时候尽情享受，因为短短儿个星期后，整个"激流勇进"将被再次拆除。

你知道吗？

最初的原木水槽是在地面开出的通道，为了把木材从山坡上运到山下的锯木厂。

1. "激流勇进"是流行的游乐设施
2. 备好防水基础，置入框架
3. 火车运来部件并一块块卸下
4. 水槽开始安装
5. 钢梁引导小车沿着水槽滑行
6. 先固定上部分的承重横梁

7. 工人在 26 米高的横梁上完成工作

8. 最后的一段水槽安装到位

9. 起重机把小车吊进来

10. 传送带把小车和乘客送到高处

11. 空车运行一天以确保工作正常

12. 搭建完成，工人们尽情地享受成果

Mobile Log Flume

扫描二维码，观看英文视频。

The log flume is one of the most popular rides at the fair. Building one is clearly a lot of hard work, and the guys who put up this flume could tell you all about it, as they have to did it nearly every month as the fair moves on from town to town.

The construction will take about a week and the first stage is to lay a base for the ride. Obviously for a log flume it needs to be water proof. They're building the largest mobile log flume in the world. And it will be filled with more than one and a half million litres of water.

The ride is moved up to ten times a year which results in wear and tear so they have to regularly service the equipment. Rips in the base are repaired with a silicon sealant. The frame work is slotted into these holes and the rubber rings will form a waterproof seal. A crane lowers down the sections, and work men guide them into place. All of the parts of the flume arrive on this train, and are lifted off piece by piece.

Whilst the framework for the top section of the ride is still being built, the first canal segments are fitted bellow. When the ride's finished there will be 52 of them in all. These steel girders will guide the carts along the canals. Twenty men work as a crew to get the whole ride up and running. And at the end of day one they've made some solid progress.

The following morning they start off by fixing the beams that will carry the top sections of the flume. These guys might work for the fun fair but they need to have the skills of circus acrobats as they balance on top of beams as high as 26 metres.

Work continues at a rapid pace and the day flies by. The last canal section is fitted into place and the waterway is complete. They use the crane to lift over the carts that will go round the flume. They are worth over five grand a piece.

The next thing they need to make a water ride work is water. To fill the system with over one and a half million litres takes almost 48 hours and costs around 5000 pounds each time.

By the end of the 5th day, the ride is up and running. But there's a long way to go before the public will be able to ride on it.

Empty carts are sent round the ride for a day to make sure there are no loose joints and the 5000 parts that make up the ride are all working as there should be. Any problems show up on the system in the control room and they're checked out by a mechanic.

The conveyor belts have to be tough enough to carry the weight of the carts and up to 6 passengers at a time. It takes them up from the bottom to a height of twenty six metres. It won't exactly be the ride of the passenger's life but it will provide some light entertainment.

10 massive water pumps dotted around the system keep the water pressure up and regulate its flow. The enormous pressure these pumps generate carries nearly 30,000 litres of water around the system every hour.

When the tests are over and the ride's deemed safe the workers finally get to have some fun. They best enjoy it while they can, because in a few short weeks they'll be taking the whole thing down again.

Did you know?

The original log flumes were channels cut into the ground which carried timber down mountainsides to sawmills below.

冰雕

扫描二维码，观看中文视频。

说说冰雕，它是一种很不寻常的爱好。一个高手能给冰赋予生命。艺术家是如何把冰块变成不可思议的雕塑呢？

从冰块转化为美妙艺术，从大量的水开始。专门从事冷冻业的公司生产了440吨晶莹剔透的冰，用作艺术家进行雕塑的原料。

让水冻起来但仍保持清澈透明是基本要求。要做到这一点，水在冰柜里须不断搅动以防止瑕疵形成。这不是一件容易的事。这么多的冰需要将近6周才能准备出来。每个大冰块都沉重得难以置信，因此运输工人搬动的时候总要非常小心。

冰雕在冬季寒冷的地方很受欢迎。这类的比赛吸引了来自中国，挪威和美国的雕塑家。每个参赛队必须小心地将冰块排好再开始工作。为做出一件优秀原创的作品，艺术家之间有很多的竞争。有些雄心勃勃雕塑家，需要大型起重设备来准备冰料。

等一切都到位，细致的工作就可以开始了。从雕刻城堡微型窗户的小锯，到切割大冰块的链锯，各种工具一应齐备。

抹刀和刮刀用来确定细节并去掉大量多余的冰块。或者像这个，做一个屋顶和城堡配套。艺术家尽量让自己的雕塑显得真实。这位正用链锯开出城堡坐落的峭壁。

冰雕的成败在于细节。收尾处理是格外精心细致的工作。这位雕刻家正把屋顶放上塔顶。他离地面6米多高，一个打滑就会让这座冰雕城堡轰然倒塌。一旦放到正确位置，他将检查是否水平并拿出水枪。胶水或水泥在这里都没有用，但两个冰块能把它们中间的水冻住，形成水泥般坚固的连接。

这样，从一堆冰块开始，变成了德国新天鹅城堡。简单的冰料变成了符合比例的路德维希二世国王的巴伐利亚童话城堡。

有时候要在雕塑上添加一块冰，就必须把它熨平。是的，你没听错，将金属板用熨斗加热后，在冰雕上融化出一个平坦的表面。然后将添加的冰块放到合适位置，再次用水把两块冰牢固地连接在一起。

刮刀和一排奇形怪状的叉子在最后造型中被用来做出不同的质感。

但真正使冰雕如此美丽的诀窍，是使用热吹风。这种好像烘干头发用的吹风机能融化冰雕粗糙的表面，再次冻结后就恢复了初始的晶莹剔透。当然，如果你没有吹风机，也可以用熨斗做出同样的效果。

当工作完成后，4个长方形大块冰就变成了一台三角钢琴。作为收尾工作的点睛之笔，艺术家用彩色灯光凸显各种细节，展示出自己作品的清澈和质地。

有些作品甚至可以和人互动，比如这个的巨大的冰滑梯就能真正玩耍。冰冷的激情，冰冻的魅力，这就是冰雕的艺术。

你知道吗？

有纪录以来最高的冰雕达12.28米，是世界上最高的商业建筑——迪拜阿拉姆塔的复制品。

1. 艺术家让冰块变成美丽的雕塑
2. 冰雕比赛吸引了各国高手参加
3. 众多优秀的原创作品展开竞争
4. 有人动用大型起重机准备冰料
5. 刮刀用来做屋顶和城堡的细节
6. 用链锯开出城堡的峭壁

7. 通过喷水把屋顶和塔顶冻牢
8. 冰块变成巴伐利亚童话城堡
9. 热铁板把冰面融平再用水冻牢

10. 热吹风或熨斗融化冰雕表面
11. 四个大冰块变成了三角钢琴
12. 冰滑梯能真正玩耍

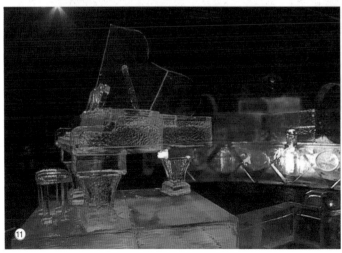

Ice Sculpture

扫描二维码，观看英文视频。

But first ice sculpture. Its one of the more unusual hobbies, and a master can bring ice to life. But how do artists turn blocks of ice into these amazing sculptures?

This transformation from ice cube to cool art starts out with water and plenty of it. Dedicated companies that specialize in freezing water produce over 440 tons of ice as crystal clear blocks for the artists to work with.

Getting the water to freeze but remain clear is essential. To do this, the freezers must continually agitate the water to stop any imperfections forming. But it's not an easy job. This much ice takes nearly 6 weeks to prepare. Each frozen block is incredibly heavy so the delivery men take great care when lining them up, most of the time.

Ice sculpting is popular wherever there are cold winters and events like these attract sculptors from China, Norway and the USA. Each team has to carefully align their blocks before they can get to work. There's a lot of competition among artists to produce an outstanding original. Some sculptures are very ambitious and need heavy lifting equipment to prepare the raw ice.

Once everything's in place, the detailed work can begin. A variety of tools are used from small saws for tiny windows in this castle, to chainsaws for carving the larger blocks.

Shaped spatulas and scrapers help define the detail and remove large lumps of unwanted ice, or in this case, craft a roof fit for a castle. The artists try to make their sculpture as realistic as possible. One of them uses a chainsaw to shape the cliffs the castle would sit on.

Ice sculpture is all in the detail. Adding the finishing touches is delicate work that needs to be handled with care. Here the sculptor is putting the roof onto the top tower. He's over 6 meters above ground … just one slip and this ice castle could come crashing down. Once he's got it in place he checks its level and pulls out his water pistol. Glue or cement won't work here, but the two frozen blocks will freeze the water in between them which create a cement-like bond.

So what started out as a pile of ice cubes, has been transformed into the Castle Neuschwanstein in Germany.

From plain ice to a scale model of King Ludwig the Second's fairy-tale Bavarian castle.

Sometimes, to add a piece to your sculpture, you need to iron it. Yes, you heard right. The iron is used to heat a metal plate which melts a flat surface onto the sculpture. The extra block is then put in place and water is used once again to create a solid bond between the two.

Scrapers and an array of odd shaped forks are also used to create different textures in the final model.

But the real trick to making ice sculpture as beautiful as it is, is the heat gun. Using this hair-dryer like device, the rough surface of carved ice melts and when it refreezes it recovers the crystal clear brilliance it started with. Of course, if you don't have a hair-dryer, you can always use your iron again to achieve the same effect.

When the work is done, 4 big rectangular blocks of ice can be transformed into a grand piano. For a finishing touch the artists use coloured lights to highlight different details and show the clarity and quality of their work.

Some of which is even interactive like this enormous ice slide which actually works. Chilled thrills, the frozen fascination that is ice sculpture.

Did you know?

The tallest ice sculpture on record was a 12.28 metre replica of the world's tallest commercial building the Burj Al Alam tower in Dubai.

DVD 生产

自从 1998 年在欧洲问世，DVD 已经席卷全世界的家庭电影。它们用满载信息的闪亮激光束，取代了老式的卷轴磁带。

生产从编辑软件开始。导演将大家喜欢的电影或者演出进行剪辑。最终的节目存储在磁带上或者计算机硬盘上，随后拿到 DVD 工厂。虽然它们通常看起来有光泽，好像金属并且反光，但 DVD 的诞生却从一片透明玻璃盘开始。玻璃盘上涂有非常薄的光敏材料层，用来储存信息。

把涂好的玻璃盘放进激光机。在这里大片电影可以写入或蚀刻在玻璃盘表面。数十亿小凹痕被激光蚀刻到感光层上。储存了应有的信息，玻璃盘现在需要变成 DVD 母版。玻璃基盘在镀镍槽中经过 1 小时的处理，凹痕表面覆盖了镍。

这些凹痕承载了所有信息，这种情况下我们的电影写入了玻璃基版。

这是个 DVD 母版，数以千计的电影或音乐视频拷贝可以由它印制出来。印制之前，模板需要被切割成合适的大小，否则将和印刷机不相匹配。

准备停当后把它装到机器里，全新的 DVD 就可以生产了。必须非常小心谨慎，因为母盘只有一张。一种被称为聚碳酸酯的塑料原料送进印刷机。熔融态的塑料压向母版表面。新的 DVD 便每 3 秒一张印制出来。

你可能已经发现，新光盘是完全透明的，这是个问题。DVD 播放机读不了这样的光盘，现在它们需要镀一层铝。接着做另一片透明的塑料光盘。遇到光碟跌落的情况，将能对镀铝层和所有重要信息起保护作用。下个机器在两者之间喷洒一层薄薄的胶水，将他们粘在一起。随后的旅程是紫外线照射。使胶水永久性固化而不会分离。

为了从这张闪亮的光盘获取图像，DVD 播放机向它发射激光束。这束激光通过透明层，被铝层反射回来。激光束反射到接收器上，再转化成你电视机的图像和声音。

现在需要识别光盘，机器首先用红外线检查是什么光盘正在播放。它们被涂上一层白色颜料，准备接下来添加其他颜色。颜料均匀涂到滚筒上，并印到每张通过的光盘上。质量控制检查员密切注视着最终完工的光盘。

如果你期待一部动作冒险电影，却碰巧拿到一部浪漫喜剧，可能会感到失望。因此，确保正确的光盘装进正确的盒子非常重要。就这样，从一张普通的玻璃盘开始，变成了一张精巧而超薄的 DVD，里面装着最新好莱坞大片。

你知道吗？

自从推出以来不到 10 年，英国人购买了超过 10 亿张 DVD，使其成为英国历史上最成功的电子存储形式。

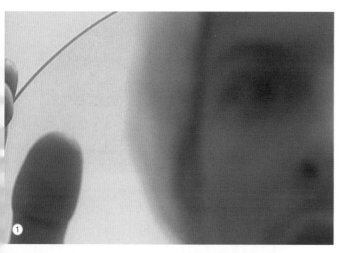

1. DVD 从一片透明玻璃盘开始
2. 涂有光敏材料的玻璃盘放入激光机
3. 影视片蚀刻在玻璃盘表面
4. 玻璃盘在镀镍槽中处理
5. DVD 母盘切割成合适大小
6. 模板装进印刷机准备生产

7. 为透明塑料光盘镀铝

8. 胶水把透明空白光盘黏合

9. 激光读出数据的原理

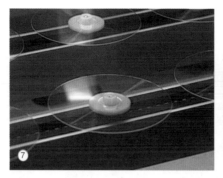

⑦

⑧

⑨

10. 机器用红外线检查播放内容

11. 滚筒的颜料印在通过的光盘上

12. 玻璃盘变成精巧的 DVD

⑩

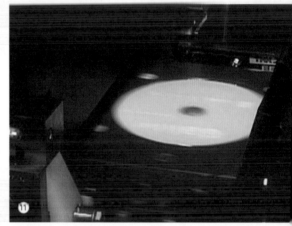

⑪

⑫

DVD Production

扫描二维码, 观看英文视频。

Ever since their introduction to Europe in 1998. DVD's have taken the home movie world by storm. They replaced old spools of a VHS tape with glittering laser beams of light chock full of information.

Production begins in an edit suite. This is where the director will cut the show or the movie that we all want to watch. The final programme can then be stored either onto a tape, or a computer hard drive which can then be taken to the DVD factory. Now although they are usually shiny, metallic and reflective, a DVD actually starts life as a disc of clear glass. This is coated with a very thin layer of photosensitive material so that information can be stored onto it.

The freshly painted disc is then put into the laser machine. This is where the blockbuster movie can be written or etched onto the surface of the glass. The laser etches billions of tiny notches into the light sensitive layer that has been painted on. With the information in place, the glass disc now needs to be turned into a DVD template. The glass master is washed in a nickel bath for an hour and the nickel collects in the notches.

These notches carry all of the information, in this case our movie, to form the glass master.

This is a DVD template and thousands of copies of the film or music video can be printed from it. Before printing, the template now needs to be cut down to size otherwise it won't fit into the printing press.

When it's ready, it's fitted into the machine and brand new DVD's can be produced. It has to be done carefully though as this is the only copy that exists. The printing press is filled with the raw material which is a form of plastic called polycarbonate. This is melted and forced against the surface of the negative. A new DVD is pressed like this every 3 seconds.

Now, you may have spotted though that these new discs are completely clear which is a problem. A DVD player can't read them like this so they now receive a coat of aluminium. This is then followed by another clear plastic disc. This will protect the aluminium layer and the all-important information in case you ever drop the disc.

The next machine sprays a fine layer of clear glue between the two to bond them together. This is then followed up by a trip under the UV light. This hardens the glue permanently and stops them coming apart.

To get a picture off this shiny disc, your DVD player shoots a laser beam at it. This beam would pass through a clear disc but now it reflected back off the aluminum layer. The laser beam bounces back onto a receiver and is translated into pictures and sound on your TV.

Now you have to identify the disc. First the machine checks which discs its working on using an infrared device. They're sent to receive a layer of white paint. This prepares them for the colours which are added in next. The paint is applied to the rollers, which spread it evenly and apply it to each disc as it passes underneath. A quality control inspector keeps a close eye on the final discs that emerge.

If you were looking forward to an action adventure and happened to find a romantic comedy, you'd probably be disappointed so it's important to make sure the right discs get into the right box. So what started out as an ordinary glass disc, has been turned into an ingenious wafer-thin copy of the latest Hollywood Blockbuster on DVD.

Did you know?

Less than 10 years since its launch, Briton have bought over 1 billion DVDs, making it the most successful electronic format in British history.

模型赛车

扫描二维码，观看中文视频。

它们排队等待出发的信号，开动！塑料观众和标记轨道清楚地显示，这是经典的 Scalectrix 模型赛车。

早在游戏机和任天堂之前，Scalectrix 就是男孩必须拥有的玩具。在今天它仍然很受欢迎，其成功秘诀的关键部分是对模型车细节的重视。设计师会花费数小时分析每一辆新的赛车，帮助他们制造出和真实车辆尽可能相似的模型车。

一旦所有研究完成后，另一个设计团队将着手为新的模型车制图。

这个公司在哪里把设计变成真正的汽车模型呢？当然是以高性价比制造业而著名的中国。这个香港周边的玩具厂，就是生产模型车竞标的胜出者。

像许多现代玩具一样，赛车模型由塑料制成。塑料颗粒被送进注塑机，熔化后形成玩具零件。当然不仅汽车需要制作，赛道也是用塑料做出来的。

这个金属造型是法拉利一级方程式赛车的模具。做出来的模型车车身是红色的，但还不是法拉利红。然而注重细节意味着每个部件都要涂成完全正确的色调。

每个模型会被喷涂上 5 层颜料以确保表层牢固，然后晾干以便进行更细致的加工。这家工厂大约有 2000 个雇员，要生产出一套完整的模型赛车玩具，需经过多达 4000 个独立步骤。

这里的大部分工作非常注重细节，必须通过手工来完成，比如把面罩玻璃画到驾驶员的头盔上。对于较大的零件，则会使用一种类似丝网印刷的技巧。只让颜料到达需要着色的特定部分，就像这条赛道的标记。

当涉及更复杂的细节时，需要更精巧的印刷方法。微型的跃马车标使用特殊的戳记印到塑料上，图案被蚀刻在印戳的顶部。当所有的细节工作完成后，汽车的比例模型应该看起来和真车几乎完全一样，包括最微小的赞助商标志。

现在外观看起来没问题了，但这些模型车也是为了参加比赛的，所以接着必须安装发动机和车轮。首先安装仿造轮胎。它们像真正的轮胎一样由橡胶制成。轮胎会帮助模型车抓住塑料轨道，就像真正的赛车轮胎一样。

F1 模型车中最重要的部分是它的发动机。虽然没有真正法拉利引擎的轰鸣声和巨大动力，却很适合在客厅的地板上比赛。

Scaletrix 赛车是通过轨道供电的。安装在汽车前部的金属接触头把能量传送到一个小电动机上使车轮旋转。在出厂之前赛车模型需要进行测试以确保轮子正常工作。你们中的一些人可能已经注意到轨道是塑料制成的，它不导电。因此就需要这些金属条。将它们插入后，把电力从电源输送到每辆车的金属接触头上。

你知道吗？

顶级直线竞速赛车模型可以比真正的 F1 赛车更快地加速。它们能在 0.2 秒内速度从 0 达到每小时 60 英里。

当整套赛车测试完毕后，就包装起来，准备发送到各地玩具店。每套玩具配有足够建造一个环形赛道的车轨，一个带有插头的电源连接器，和两个不同的赛车模型，可以用来进行那些并驾齐驱的精彩比赛。

通过真正的货车，而不是货车模型，套装玩具随后被运到码头，开始前往世界各国的漫长旅程。因此无论你和朋友比赛，或只想打破自己的圈速纪录，赛车模型都是一种令人兴奋的游戏，让你在自己家中体验赛道上的惊险刺激。

①

1. Scalectrix 赛车备受男孩欢迎
2. Scalectrix 成功的秘诀是细节
3. 设计师仔细分析新的赛车
4. 模型赛车用塑料制作
5. 这是法拉利一级方程式赛车模具
6. 模型被喷涂上 5 层颜料晾干

②

③

④

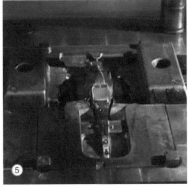

⑤

⑥

7. 大零件用类似丝网印刷术
8. 专用车标外观和真车几乎一样
9. 安装的仿造轮胎由橡胶制成
10. 电动机通过轨道获得电流
11. 电通过金属触头开动模型车
12. 赛车模型和配件装盒

⑦

⑧

⑨

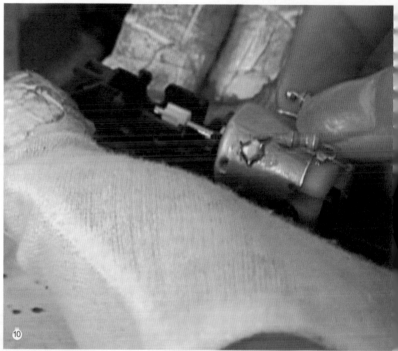

⑩

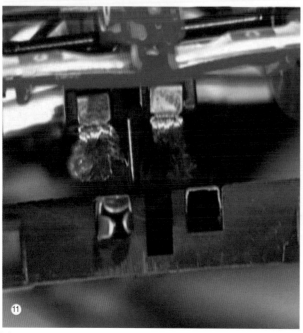

⑪

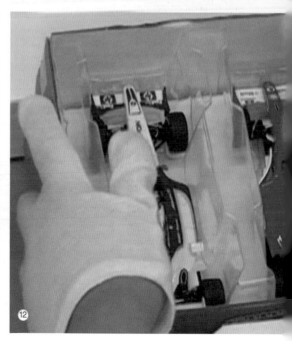

⑫

Model Cars

But first, they're lined up just waiting for the starting signal, and they're off! The plastic spectators and tell-tale track make it clear that this is classic Scalectrix model car racing.

Long before Playstations and Nintendo, Scalectrix was a must have toy for boys. It's still popular today, and a key part of its winning formula is the attention to detail on the model cars. Designers will spend hours, analysing each new race car to help them create a model that looks as much like the real thing as possible.

Once all the research has been done, another design team will go about drawing up plans for a new model car.

And where does the company transform its designs into real replicas? Well it's China of course, famous for it's cost effective production. This toy factory outside Hong Kong is where the race to make model cars will be won.

Like many modern toys, replica models are made out of plastic. Granules are fed into injection moulding machines to be melted and shaped into the toy parts. Of course it's not only the cars that need to be crafted. The race track is also made out of the plastic.

This metal shape is the mould for a Formula 1 Ferrari. The model-car bodies that emerge are red, but not Ferrari-red. However, attention to detail means that each part gets coated in exactly the right shade.

Each model will be sprayed with 5 layers of the paint to ensure a solid surface. They're then left to dry before the more detailed work is done. Around 2000 staff work at this factory and there can be anything up to 4000 separate steps involved to create a full model race set.

Much of the work here is so detailed it must be done by hand, like painting a visor onto the driver's helmet. For the larger parts a technique similar to silk screening is used. The screen only let paint through to specific parts that need colouring, like the markings on this piece of track.

When it comes to the more intricate details a finer printing method is needed. The miniature prancing stallion logo is printed onto the plastic using special stamps that have the design etched on the tips. When all of the detailed work is complete, the scale model should look almost exactly like the real thing, down to the tiniest sponsor's logo.

Now, its all very well getting the look right, but these model cars are meant to race too, so the engine and wheels must be attached next. First the imitation tyres are fitted. Like the real thing they're made of rubber. They'll help the model to grip the plastic tracks just like the tyres on a real racing car.

And the most important part of the F1 model is its engine. While it won't have the roar or raw power of a real Ferrari motor, it will be fit for racing on a living room floor.

Scaletrix cars are powered by electricity from the tracks. Metal pick ups attached to the front of the car feed power to a small generator that spins the wheels. Before they leave the factory the models are tested to make sure the wheels are working properly. Now some of you might have noticed the tracks are made of plastic, which doesn't conduct electricity. That's what these metal strips are for. They're inserted to carry electricity from the power supply to the metal-picks ups on each car.

Once the full racing set has been tried and tested, it's packaged up ready to be sent out to toy stores everywhere. Each kit contains enough track to build a looping course, a power connecter with a handy plug, and two different replica cars for some great head to head racing.

Using a real lorry, not a replica one, the kits are then taken to the docks for their long journey to countries all over the world. So whether you're racing your friends or just trying to beat your own lap record, replica racing cars are an exhilarating way to experience the thrills of the race track in your own home.

Did you know?

Top drag racing model cars can accelerate faster than a real F1 race car. They can do 0 ~ 60 miles an hour in 0.2 of a second.

日本太鼓

扫描二维码，观看中文视频。

日本以现代创新科技而著名，也是一个传统文化和礼仪丰富的国家。太鼓手在适应新时代的同时，体现了对旧传统的敬慕。打太鼓需要力量和技巧，它们的声音闻名世界各地。但不只是打鼓需要技巧。日本早在 2000 年前便制造出了太鼓。生产这些艺术品乐器的工艺经过很多世纪流传下来。

传统上，鼓由榉木制造。榉木是一种生长在日本山区的坚硬木材。首先必须让木头干燥，在能够进行加工之前，干燥需要大约 1 年时间。

鼓匠截下一段约 1 米长的木头来制作一面普通大小的太鼓。最早需要用手工锯断木头，如今有现代化的链锯来做这项艰苦工作，一些传统也就让路了。使用榉木的另一个原因是它精美的纹理。一面太鼓最终的外观和它的声音同等重要。

现在鼓匠在树干上标出鼓的形状。手工切掉多余的木材是件非常艰巨的工作，但现代工具再次节省了大量的时间和精力。鼓匠不但要去掉难看的树皮还要切割出新鼓的内腔。当完工后，他得到成品乐器的鼓身和插在里头的一个大塞子。这时候巧匠的精细手工接替了电锯的粗糙切割。

一旦准确的形状由手工制作完成后，鼓身被留在一个这样的房间慢慢变得干燥和稳固。室温为恒定的 25 摄氏度，这个过程可能花费多年时间，取决于新鼓的大小。像这样的大鼓重约 800 千克，需要放置 12 年才能最终加工为成品。最大的那些鼓造价超过 28 万英镑。

当鼓身准备就绪，将进入生产的下一个工序。这个木匠正在制作太鼓令人惊叹和充满活力的外观。他将外表面打磨光滑，更突出了榉木显眼的天然纹理。

为了让太鼓具有传统的光彩和色泽，他会给每个鼓身涂上几层透明清漆，这样既保护了木材又赋予它更厚重的色调。

但除了木制鼓身，太鼓能够敲响还需要其他部分。鼓皮的生涯中注定要承受数千次敲击，因此必须非常结实。皮革是鼓皮的首选，它被精心地备好，边缘带有精细的小卷曲。稍后你会知道它们能有什么用。

首先将鼓皮湿润，然后开始艰难的拉伸过程。鼓皮上使用润滑剂，让它具有延展性，更容易拉抻到位。鼓匠用金属销穿过我们前面看到的那些皮革上的卷曲。这样就做出了固定点，能使鼓皮绷紧。将绳子绕在金属销和放置太鼓的拉抻装置上，但这仅仅是工序的第一部分。

在确认他的金属销仍在插紧状态后，鼓匠用这些手柄将绳子绞得更紧。但即便这样也还不够。他还将使用太鼓下面基座中的液压装置。

当确信皮革鼓皮够紧了，他会测试一下，但不是用棍子。而是用他的脚。为证实皮革和

你知道吗？

纽约市的彼得·拉文格是世界上最大的鼓槌收藏家。他有 1300 个著名鼓手的鼓槌，其中许多都有签名。

木制鼓体的结实，鼓被当作一个蹦床。这有助于鼓皮移动并稳定到一个好位置。也给了鼓匠测试声音的机会。如果哪些地方不对，就需要把鼓皮拉抻得更紧。只有当鼓匠满意了，最后的音色完美了，他才会标记下金属销的位置，并开始把鼓皮固定到位。

为确保鼓皮丝毫不差地保留在被拉抻到的位置，鼓匠使用不是一排，而是两排的钉子来固定。只有当他确信鼓皮被永久固定到正确的位置时才会撤掉拉力。

太鼓最初用来在战斗中给敌方士兵制造恐惧和传递新命令。现代的太鼓通过来自日本历史的威风节拍，标志着一种古老传统和文化的遗存。这就是传统的太鼓。

1. 日本太鼓表演
2. 将干燥 1 年的榉木锯下 1 米小段
3. 沿划好的圆面去掉多余部分
4. 链锯切割出鼓的内腔
5. 手工精细挖出鼓的内腔
6. 推进恒温室

7. 在 25 摄氏度恒温下干燥多年

8. 鼓身外表打磨光滑并喷上清漆

9. 鼓皮浸湿并用绳子拉抻

10. 用液压装置将绳子拉紧

11. 蹦跳测试鼓皮松紧和音色

12. 太鼓就此诞生

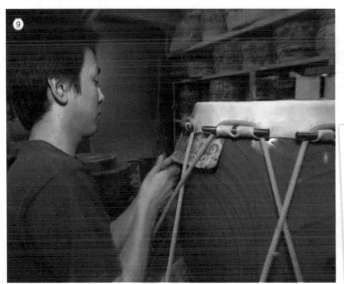

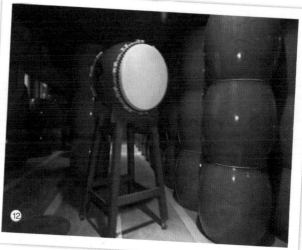

Japanese Drums

扫描二维码，观看英文视频。

Famous for its modern innovation and technology, Japan is also a nation with a rich culture of tradition and ritual. Taiko drummers embody this respect for the old way, whilst adapting to the new. It takes strength and skill to play these drums and their sound is famous around the world. But it's not just playing the drums that takes skill. Taiko drum's were first produced in Japan over 2000 years ago. The skills to craft these artistic musical instruments have been handed down through centuries.

Traditionally, the drums are made out of keyaki wood. It's a hard wood that grows in the mountainous regions of Japan. It must be dried first and this takes about 1 year before it is ready to be worked with.

The drum maker will cut a piece about 1 metre in length to create an average sized instrument. Originally this would have been cut by hand, but some traditions give way when a modern chainsaw is available to help with the hard work. Another reason for using the keyaki wood is its fine grain. A Taiko drums final appearance is as important as its sound.

The drum maker now marks out the drum's shape in the trunk. Cutting away the excess wood by hand would be a very tough job, but once again modern tools save a lot of time and effort. As well as removing the unsightly bark, he will also carve out the inner body of the new drum. When he's finished, he's left with a large plug inside what will become the body of the finished instrument, and here's where the fine handiwork of a skilled craftsman takes over from the crude cutting of a chainsaw.

Once the exact shape has been hand-crafted, the drum body is left in a room like this one to dry and settle very slowly. At a steady 25 degrees Celsius that can take many years, depending on the size of the new drum. A large drum like this weighs around 800 kilograms and needs to rest for 12 years before it can be finished. The biggest ones can cost more than £280,000.

When the drum body is ready, it's taken to the next stage of production. This carpenter is working to produce the stunning and vibrant façade of the Taiko drum. His work of sanding and smoothing the outer surface highlights the impressive natural grain of the keyaki wood.

To give the drum its traditional shine and lustre, he will give each body several coats of a clear varnish this both protects the wood and gives it a richer, deeper hue.

But there's more to the beat of this drum than just the wooden body. The skin must withstand many thousands of beats during its lifetime, so only the toughest will do. Leather is the skin of choice and this is carefully prepared with delicate little curls around its edges. You'll see what they're for shortly.

First the skins are dampened and now the tough stretching process can begin. Lubricant is added to the skin to make it malleable and easier to be pulled into position. The drum maker then passes metal pegs through those curls in the leather we saw earlier. These create anchor points so the skin can be stretched tight. Ropes are woven round the pegs and the stretching device the drum is resting on, but this is only the first part of this process.

Ensuring his pegs are still tightly fitted, the drum maker winds the ropes even tighter with these handles, but even that isn't enough. He will also use the hydraulics in the base below the drum.

When he's sure the leather skin is tight enough, he'll try it out, but not with sticks. He will use his feet. Testament to the toughness of the leather and the wooden body, the drum is used like a trampoline. This helps the skin to move and settle into a good position. It also gives the drum maker the chance to gauge the sound. If it's not right, more tightening is called for. Only when he's satisfied, the final note is perfect, will he mark out the position for the pegs and begin fastening the skin into place.

To make sure the skin remains exactly where it has been stretched to, he will use not 1, but 2 rows of nails. And he will only release the tension when he is sure the skin is permanently fixed into the correct position.

Originally used in battle to drum fear into enemy soldiers and signal new orders, the modern Taiko drum now signals the survival of an ancient tradition and culture, through an awe-inspiring beat that has its roots in Japanese history. The traditional Taiko drum.

Did you know?

Peter Lavinger, of New York city, has the world's largest collection of 1,300 famous drummers' drumsticks. Many of them are autographed.